WITHDRAWN

LASERS THE LIGHT FANTASTIC

SECOND EDITION

LASERS THE LIGHT FANTASTIC

SECOND EDITION

Clayton L. Hallmark
and Delton T. Horn

TAB BOOKS Inc.
Blue Ridge Summit, PA 17214

SECOND EDITION

FIRST PRINTING

Printed in the United States of America

Library of Congress Cataloging in Publication Data

Hallmark, Clayton L.
Lasers—the light fantastic.

Includes index.
1. Lasers. I. Horn, Delton T. II. Title.
TA1675.H35 1987 621.36'6 87-10182
ISBN 0-8306-0305-0
ISBN 0-8306-2905-X (pbk.)

Questions regarding the content of this book should be addressed to:

Reader Inquiry Branch
Editorial Department
TAB BOOKS Inc.
P.O. Box 40
Blue Ridge Summit, PA 17214

Cover photograph courtesy of Coherent, Inc., Palo Alto, CA.

Contents

Preface to the Second Edition

For this second edition, greater emphasis has been placed on laser applications. In the last few years, lasers have become increasingly important in civilian, as well as military, applications. Newer lasers are smaller, more efficient, and more precise than their predecessors.

The structure of the book is essentially the same as before, but new material (including some completely new chapters) has been added to bring the book up to date with the latest developments in the field of lasers.

Delton T. Horn

Introduction

The summer of 1960 saw the world's first demonstration of an entirely new source of light so concentrated and powerful that it could produce power densities millions of times as intense as those on the surface of the sun, yet be controlled so precisely that surgeons could use it to perform delicate operations on the human eye. The beam from this device could burn holes in steel plates and set carbon on fire. It spread out so little that if sent from the earth to the moon, it would illuminate an area of the moon's surface less than two miles in diameter. This new device was the laser. ***Laser*** means ***light amplification by stimulated emission of radiation***, the process that takes place inside the device.

Simply stated, the laser is a beam of light. The light from a house light spreads in all directions in a number of various frequencies and waves. The very fact that light does travel in waves led to the development of the laser, which concentrates the waves of light into light beams of tremendous energy.

Originally developed by the use of ruby crystals, lasers are now produced by many solid materials, liquids, gases, and semiconductor devices.

The laser is one of the most important developments of modern science and is used extensively in medicine, by the armed forces, and in every facet of industry where precise measurements are needed. Hundreds of new applications are being proposed and developed by engineers and scientists, and new uses appear unlimited in scope.

The armed forces have been the largest buyers of the laser. The more important military applications include proximity fuses, night-time vision and tracking, target illumination, and range-finding units that provide accurate measurements.

Since lasers have such a large area of actual and potential application, a knowledge of lasers and related devices is important to a wide range of persons. The potential readers of this volume have a wide range of interests and backgrounds and come from a variety of fields. To meet the needs of as many of these persons as possible, I have strived to make this a completely self-contained, self-sufficient text.

The early chapters of the book discuss general matters and provide the background for understanding the later chapters. The later chapters discuss the theory of lasers and related devices in more detail than in any popular book in print. Only large engineering texts provide greater technical detail. Traditionally quantum mechanics has been regarded as the epitome of difficult subjects. But it doesn't have to be difficult—if one has the proper background for understanding it. Since some readers may be nontechnical persons, unfamiliar with scientific measurements, Chapter 2 discusses measurements, with a view toward helping the reader understand how lasers are measured and how their specifications are stated. Since some other readers may be technical persons with a background in some field other than electronics, Chapters 3 and 6 introduce electronics principles common to electromagnetic generators, including lasers and masers. Electronics personnel may wish to read these chapters as a review, or they may wish to just skim over the chapters to see what is there, for possible future reference.

Subsequent chapters present the theory of quantum electronics and the theory of specific quantum-electronic devices. The final chapters present some applications of lasers in communication, radar, gyroscopes, space, industry, and commerce.

Chapter 12 offers guidance for the reader wanting to actually experiment with lasers, including an actual schematic and information on how to obtain the necessary materials at low cost. Before you do any experimenting, be sure to read the information on laser safety in Chapter 20. This information is useful for anyone working with lasers.

I hope you enjoy reading about, learning about, and possibly experimenting with lasers—the light of the 21st century.

Clayton L. Hallmark

Brief History of Lasers

The basic theory of lasers can be traced as far back as 1917 to atomic theories by Einstein, who pointed out that controlled radiation could be obtained from an atom under certain conditions.

Until the 1950s, lasers were strictly theory, but many scientists were working for reality. In June 1954, Dr. Charles H. Townes, then of Bell Laboratories, demonstrated an ammonia gas maser with adequate stimulation to sustain a maser beam of a signal equal to 24,000 megahertz. The real significance of this demonstration was that Dr. Townes generated a signal without the use of a resonant cavity. This achievement opened the door to many experiments leading to the amplification and generation of electromagnetic radiation within the visible light spectrum wavelength.

Laser history actually begins with the maser (*microwave amplification by stimulated emission of radiation*). Early masers were gaseous ammonia or rubidium. A maser was built in 1955, and a solid-state version was proposed in 1956. One of the first solid-state masers employed sapphire crystals (synthetic) doped with chromium oxide. It was later discovered, in 1959, that masers could amplify light as well as microwaves, although the extension of maser principles to permit operation in the light spectrum was suggested as early as 1953. This development was referred to as *optical maser*, with the name changed to *laser* in 1965.

The light generating capability became more important than the amplifying ability. The pulsed ruby laser, built in 1960, was the first to demonstrate the monochromatic and coherent light principle.

This historic prototype was developed by Dr. Theodore Maiman of Hughes Research Laboratories. Most other researchers were still ironing out the theory, writing technical papers and debating over the best way to construct a practical laser.

Dr. Maiman deserves credit for building the world's first practical laser almost on his own. Management at Hughes Research Laboratories was doubtful about his chances for success, and provided only limited funding.

A relatively simple solid-state design was used in Maiman's device. This type of laser is known as ruby laser, because it uses a synthetic sapphire crystal doped with chromium (ruby). In chemical terms, this is CR_2O_3 in Al_2O_3.

Dr. Maiman started with a rod that was about one centimeter in diameter and two centimeters long. It was machined to optical tolerance and the two ends were cut parallel to each other. The ends were then polished and silvered to enhance reflection. The rod was mounted in a coiled xenon flashlamp.

In operation, a capacitor was discharged through the flashlamp, causing it to briefly bathe the crystal in a brilliant burst of light. This energy was used to power the lasing action of the crystal. A brief pulse of extremely intense red light was then emitted through a small aperture in one of the rod's end-mirrors.

This stimulated emission of coherent and monochromatic light was at 6943 angstroms. The monochromaticity was about five times narrower than that of the ruby's natural fluorescence. Later developments brought this figure to 40 times, then to 10,000 times the natural ruby fluorescence. Still greater monochromaticity is realized with the helium-neon gas laser.

While Dr. Maiman was developing his ruby laser, other researchers were trying out other materials. These devices were more complex than Maiman's simple ruby laser, so it's not surprising that they took a little longer to build. In 1961, Bell Laboratories developed the first continuously operating gas laser. It employed a mixture of helium and neon gas excited by an rf field, and it obtained an output in the infrared range. Laser action was produced with trivalent uranium in calcium fluoride in 1960. This was followed by other solid-state experiments. During 1961, amplification of light was performed using a ruby laser driven by a ruby laser oscillator. The generation of optical harmonics was achieved in 1961 using a laser source directed into crystalline quartz. Ruby laser light at 6943 angstroms was raised to a shorter wavelength value of 3472 angstroms.

The function of junction electroluminescence in science and technology changed rapidly back in the early 1960s. In 1961 gas lasers and optically pumped solid lasers stimulated research into the possibility of a semiconductor junction laser. Then, the discovery in 1961 and 1962 that the efficiency of recombination radiation was very high in gallium arsenide focused attention on this compound as a potential laser material. The first injection lasers were operated in late 1962. In the following years, great advancements were made in the development of basic materials and in the application of these materials in workable laser devices for a wide variety of uses. Today, the principal types of lasers include solid-state lasers (for both pulsed and continuous wave operation); injection lasers (for both pulsed and CW operation); gas lasers (for both pulsed and CW operation); and liquid, plastic, and Raman lasers. Each type uses a large variety of different materials. Some materials have been standardized in working laser devices, while other materials are used only for research effects. Development is still progressing at a very fast rate with new materials appearing constantly.

Since the first practical lasers appeared in the early 1960s, literally hundreds of substances have been employed in numerous types of lasers. A fairly recent development is the semiconductor laser which offers many advantages, especially in terms of size, reliability, and efficiency.

Like many inventions, the laser was initially just a scientific curiosity. Once the device existed, however, countless applications were soon found for it. The first applications were for military purposes, but civilian use of the laser is becoming increasingly common. Lasers are used in weather forecasting, medical procedures, industrial construction, and even home stereos and video systems. New applications for the laser are constantly being found. It is not just hyperbole to call the laser the "light of the 21st century."

Matter and Measurements

The field of masers and lasers (optical masers), like practically every field of science, deals with the structure and measurement of matter. In masers and lasers, molecules and atoms are made to absorb and emit radiation. Molecules and atoms can be regarded as machines for generating microwave and optical radiation. To understand masers and lasers, one must understand these elemental machines. This chapter provides an introduction to molecules and atoms. Later chapters expand on the ideas presented here.

Lord Kelvin, one of England's great scientists, said that unless a person can describe the topic of this study with measurements, he knows nothing about that topic. It is fitting, then, to begin a study of the topic of lasers and masers with a consideration of the basics of measurements. It is particularly important to understand the metric system. Terms such as *gigahertz* and *joule*, which are frequently used in laser specifications, come from the metric system.

THE IMPORTANCE OF MEASUREMENT

Without realizing it, practically every act of our daily lives involves measurement of some kind. In driving to work or school, our speed is measured in miles per hour. Upon arriving at our destination, we determine whether or not we are late by looking at a clock—the instrument of time measurement. How handicapped the chemist would be if he could not measure the myriad quantities of chemicals in his experiments. Picture the chaotic results of a carpenter's efforts if he could not measure the length of the timbers used in constructing a home. Since measurement is the very foundation of the study of electricity and electronics, it will be discussed next.

Mechanics of Measurement

Basically all measurements, regardless of type, are accomplished in the same way. To make a measurement, one must compare the dimensions of the quantity to be measured with the dimensions of a known standard quantity. The quantity to be measured is then said to be so many times larger or smaller than the known standard quantity.

Any measurement can be divided into two parts. The first part tells the standard or reference used for comparison. As an example, assume that a halfback runs the length of a football field in ten seconds. From this measurement, his running time is determined to be ten times greater than the standard or reference—one second. This comparison of the halfback's time to the time of one second tells how long it takes him to run the length of the field.

To make measurements meaningful to other people, a reference must be used which has one exact meaning to all who use it. In the above example, everyone concerned with the measurement must agree on the length of time of one second, otherwise ten seconds would indicate a different amount of time to each person.

The Metric System

Let us now examine a particular system of measurement, namely, the metric system. This system is used by most European countries and by scientists throughout the world. Any system used for the measurement of matter is ultimately based on three basic dimensions: length, mass, and time. Scientists use a form of the metric system known as the meter-kilogram-second (mks) system in which the unit of length is the meter, the unit of mass is the kilogram, and the unit of time is the second.

The meter was originally defined as one ten-millionth of the distance along a meridian extending from the North Pole, through Paris, to the equator. At the time of this definition a physical model was created and housed at the International Bureau of Weights and Measures, Sevres, France. The physical standard is made of a platinum-iridium alloy bar upon which two marks are etched. The distance between the two marks, when the bar is at the temperature of melting ice, is one meter. Modern investigation shows that this physical model varies slightly from time to time and is therefore not truly accurate. In the early 1960s, scientists adopted a new and unvarying standard based on the length of one wave of a specific kind of light energy.

The kilogram uses as its standard a block of platinum-iridium alloy called the international prototype kilogram also preserved at Sevres, France. The kilogram is based on the gram (1/1000 of a kilogram), originally defined as: the mass of one cubic centimeter of pure water measured at 4° Celsius (C). This temperature was chosen because near 4° C the density of water is practically independent of temperature.

The second was defined as 1/86,400 of a mean solar day, the mean solar day being the average time required for the earth to make one rotation on its axis while revolving about the sun. In recent times, it has been found that this standard also has irregularities, and in the early 1960s an atomic clock based on the vibration of a cesium atom was adopted.

A subdivision of the metric system is known as the centimeter-gram-second (cgs) system, in which submultiples of the meter and kilogram are used. A centimeter is one-hundredth of a meter; the gram and second were explained previously.

Convenience Units

In many situations the quantities to be measured can be extremely large or extremely small, and the basic units can prove too cumbersome for ease of manipulation. For example, the length of our galaxy, the Milky Way, is approximately (946,800,000,000,000,000,000) meters. It should also be kept in mind that, compared to other astronomical distances, the length of the Milky Way is small. Thus, to the astronomer the unit meter is not convenient. A much more meaningful description of this distance would be 100,000 light years, where the light year is defined as the distance light travels in one year.

In the metric system a prefix is attached to one of the basic units to provide a unit more consistent with the dimensions of the quantity involved. A list of the commonly used prefixes and their meanings is given in Table 2-1. A common example of the use of prefixes is the microsecond, which is one-millionth of a second.

British Gravitational System

A second system, the one common to this country and Britain, also deserves mention. This system, known as the foot-pound-second (fps) system is the one with which the average person in this

Table 2-1. Metric Prefixes.

PREFIX	ABBR	POWER OF TEN	VALUE
TERA	T	10^{12}	MILLION MILLION
GIGA	G	10^{9}	THOUSAND MILLION
MEGA	M	10^{6}	MILLION
KILO	k	10^{3}	THOUSAND
HECTO	h	10^{2}	HUNDRED
DECA	dk	10^{1}	TEN
-	-	10^{0}	ONE
DECI	d	10^{-1}	TENTHS
CENTI	c	10^{-2}	HUNDREDTHS
MILLI	m	10^{-3}	THOUSANDTHS
MICRO	μ	10^{-6}	MILLIONTHS
NANO	n	10^{-9}	THOUSAND MILLIONTHS
PICO	p	10^{-12}	MILLION MILLIONTHS
FEMTO	f	10^{-15}	-
ATTO	a	10^{-18}	-

country is most familiar. The fps system differs from the metric system in that the pound is not a unit of mass as is the kilogram, but is a unit of force. The unit of mass in the fps system is the slug, which is equal to approximately 14.6 kilograms. Formerly, standards were maintained for the fps system; but now all these units are defined in terms of metric standards. The relationships between metric and British units are shown in Table 2-2.

Measurement of Light

The study of light has provided man with a fascinating but most perplexing problem in science. Man has learned to generate, control, and measure light energy very effectively, even to amplify it in lasers. Through measurements of the faint light coming from distant

Table 2-2. Comparison of Units.

METRIC TO BRITISH	BRITISH TO METRIC
1km = 0.62137 mi	1 mi = 1.6093 km
1m = 3.2808 ft	1 ft = 0.3048 m
1 cm = 0.3937 in.	1 in. = 2.5400 cm
1 kg = 2.2046 lb	1 lb = 0.4536 kg
1 gm = 0.0353 oz	1 oz = 28.3490 gm

stars and planets man has learned practically all that is now known about the objects in outer space. Unfortunately the exact structure of light is still a mystery. It is well known that light is a form of energy, but the physical form in which this energy exists is not known. Light appears to be simultaneously both an electromagnetic wave and a stream of massless energy particles called photons.

One of the most important measurements associated with light energy is its wavelength. Light can be analyzed by assuming it consists of waves similar to the ripples generated when a ball is dropped into a pool of water, as illustrated in Fig. 2-1. The waves which are generated consist of a number of *cycles* such as the one shown between points *A* and B. In traveling from *A* to *B* the wave has gone through all of its possible variations and therefore has completed an entire cycle of events. In traveling from *B* to *C* the wave would simply repeat the variations that occurred between *A* and *B*. The number of these complete cycles per second (hertz) is called the ***frequency*** of the wave. If the wave illustrated completes one cycle in 1/20 of a second, it would have a frequency of 20 cycles per second.

The distance between a point on a wave (*A*) and a corresponding point on an adjacent wave (*B*) is called the ***wavelength*** of the wave. Wavelength can be measured in any of the distance units described previously such as inches, feet, meters, or centimeters, etc. Light waves have extremely short wavelengths of less than one-millionth of an inch.

In dealing with lasers and masers, we are more interested in wavelength than in frequency. The very small wavelengths of lasers and masers are expressed in terms of ***microns*** (μ), ***nanometers*** (nm) and *angstroms* (Å). The relationships among these units are as follows:

1 micron:

1/1,000,000m (10^{-6}m)
1000 nm
10,000Å

1 nanometer:

1/1,000,000,000m (10^{-9}m)
1/1000μ
10Å

1 angstrom unit:

1/10,000,000,000m (10^{-10}m)
1/10,000μ
1/10 nm

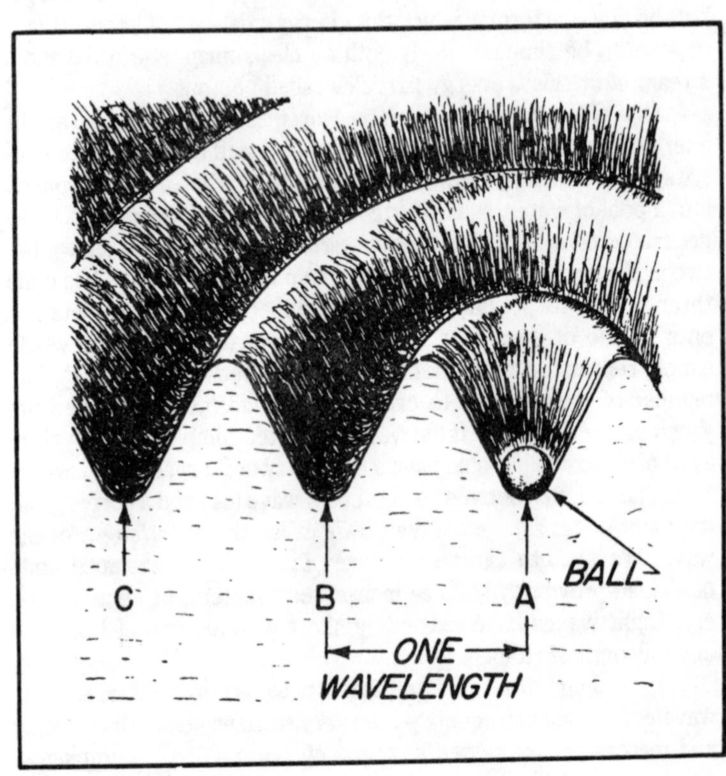

Fig. 2-1. Concept of wavelength.

Light waves vary in length from 3.85 to 7.60 ten-thousandths of a millimeter, which is 0.385 to 0.760 microns, 385 to 760 nanometers, or 3850 to 7600 Angstrom units.

In the metric system, angular measurements are expressed in terms of ***radians*** and submultiples of radians rather than n degrees, minutes, and seconds. A radian is 57.296 degrees. The very slight divergence of laser beams is expressed in ***milliradians*** (mrad), or thousandths of a radian. The equivalence between milliradians and U.S. customary units is as follows:

1 milliradian:
0.057296 deg
3.438 min
206.265 sec of arc

Measurement of Heat

Heat is a form of radiant energy very similar in nature to light. For many purposes, heat energy (often called thermal energy) can be considered to be light energy of a wavelength too long for detection by the human eye, that is, infrared.

The quantity of heat a substance contains is normally measured in one of the following basic units: a *calorie*, which is the quantity of heat necessary to raise the temperature of one gram of water one degree Celsius, or the ***British thermal unit*** (Btu), which is the quantity of heat necessary to raise the temperature of one pound of water one degree Fahrenheit.

From the above definitions it can be determined that the temperature of an object is by no means a measure of the amount of heat it contains. A small quantity of water in a test tube can be raised a degree in temperature by the heat from a wooden match. In contrast, consider the great amount of heat required to raise the temperature of the water in a swimming pool one degree. It is evident that these two quantities of water, while they may be at the same temperature, contain vastly different quantities of heat. Temperature indicates the extent to which a body has been heated, rather than the amount of heat which it contains. Although many scales have been designed for the measurement of temperature, only the Celsius (formerly called centigrade scale) and the Fahrenheit scales will be discussed here. The main differences between the two scales illustrated in Fig. 2-2 are the values arbitrarily assigned to the freezing and boiling points of water. On the Celsius scale the freezing and boiling temperatures

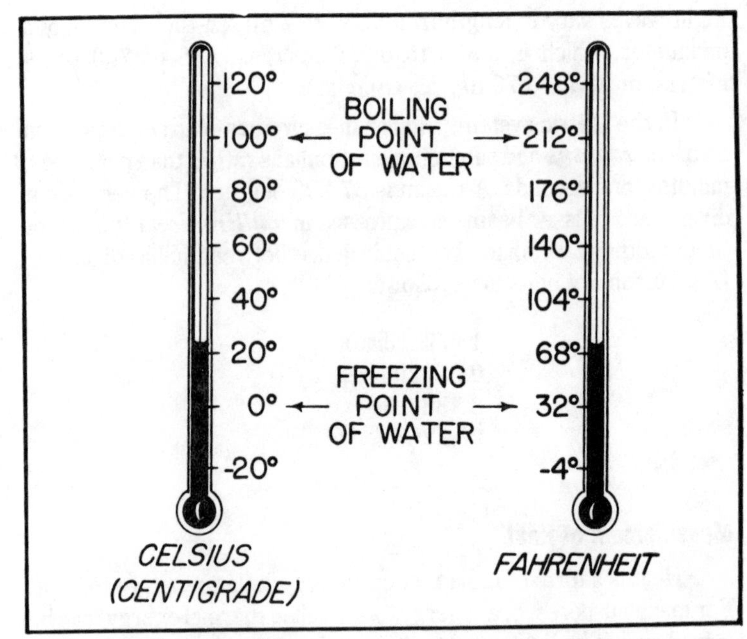

Fig. 2-2. Temperature scales.

are 0 degrees and 100 degrees respectively, while on the Fahrenheit scale they are 32 degrees and 212 degrees respectively. Equations for conversion from Fahrenheit to Celsius or from Celsius to Fahrenheit can be found in most physics books and will not be presented here.

Measurement of Force

Although the word force usually causes a person to form a mental image of some type of mechanical system, the origin of many forces is electrical in nature. The operation of the electrical motors in vacuum cleaners, refrigerators, and other home appliances is entirely dependent upon the interaction of electrical forces.

Force is defined as *that quantity which causes acceleration (change in motion) of a material body*. The pound, a unit with which we are all familiar, is the unit of force in the British system. A pound is the amount of force necessary to impart an acceleration of one foot per second per second of a mass of one slug.

In the mks system the force unit is the newton (NP) and is defined as *that force which will give a mass of one kilogram an acceleration of one meter per second per second*. A newton would be the approximate force felt on one's hand while holding a one-quarter pound package of butter.

In the cgs system the unit of force is the dyne (dyn). One dyne is the amount of force that will give a mass of one gram an acceleration of one centimeter per second per second. A dime coin held on the finger tips would exert a downward force of about 2450 dynes. This example indicates that the dyne is a rather small unit of force.

Measurement of Pressure

Another quantity which acts on matter is pressure. Pressure is defined as *force per unit area*, and is expressed as a force unit divided by an area unit. Thus, pressure can be expressed as pounds per square inch (psi), dynes per square centimeter, etc.

The atmosphere surrounding the earth exerts a pressure of approximately 14.7 pounds per square inch on the surface of an object placed at sea level. This normal sea level pressure is sometimes used as a unit of pressure, called an atmosphere (atm). One atmosphere is therefore a unit of pressure equal to approximately 14.7 pounds per square inch. Pressure can also be measured in bars or microbars. One bar is equal to 14.5 pounds per square inch or one million dynes per square centimeter.

Pressure can also be measured by comparing it to the pressure exerted by a column of mercury. Thus, a pressure can be stated as seven millimeters (mm) of mercury (Hg) indicating a pressure equal to that exerted by a column of mercury seven millimeters high. Similarly, a pressure of ten centimeters (cm) of mercury would be the pressure exerted by a column of mercury ten centimeters high. Over the years many different units have been devised for the measurement of pressure. However, the ones listed above are the most common.

CHARACTERISTICS OF LASER MATERIALS

Matter is known to exist in three states: gas, liquid, and solid. Matter in the gaseous state will conform perfectly to the shape of its container. It possesses neither a fixed shape nor a fixed volume. Some common examples of gases are the atmosphere which we breathe, the carbon dioxide which we exhale, and water vapor. Gases used in lasers include carbon dioxide, helium, argon, xenon, and ammonia.

A liquid differs from a gas in that it has a fixed volume. It is similar to a gas in the respect that it has no fixed shape, and will conform to the shape of its container. Examples of common liquids are water, petroleum, and mercury. Laser action has been observed in some special liquids.

In a solid state, matter has a fixed shape and volume. Common solids are iron, quartz, and carbon. Solids are frequently used in lasers. Ruby crystal is an example. Of the action takes place at the junction of two semiconductor materials, as in a gallium-arsenide (GaAs) diode, the device is termed a semiconductor diode laser.

Change of State

One of the fundamental properties of matter is its ability to change state. A change of state is most conveniently observed in the substance we know as water. We are all familiar with fact that water, which is normally a liquid, is easily converted to a solid, ice, or to a gas, steam. A moment's consideration of the three states of water indicates that some external influence must be involved in producing a change of state. Extending our reasoning along this line, the first thing that comes to mind is that steam is very hot and ice is very cold. From this, one may correctly assume that heat is one of the factors involved in a change of state. Water at a temperature of 0°F is solid ice. If the temperature is raised above 32°F the solid ice becomes the liquid form of water. If the temperature is raised still higher to 212°F, the liquid vaporizes into the gas known as steam.

A second factor involved in a change of state is pressure. If the temperature and pressure of oxygen are adjusted to the proper value, it can be made to exist as a liquid. At greater pressure and reduced temperature it will be converted to a solid. Since the properties of many substances are dependent on both temperature and pressure, a standard or reference is necessary. This reference condition is known as standard temperature and pressure (STP), or normal temperature and pressure (NTP), and is defined as a pressure of one atmosphere at a temperature of 0°C.

Some materials, such as dry ice and moth balls, change from the solid to the gaseous state without going through the liquid state. These materials are called *sublime* materials and the process by which this takes place is called *sublimation*. Sublimation can occur even in metals and other hard substances exposed to lasers, which aids in drilling with lasers.

INTERNAL STRUCTURE OF MATTER

With the exception of the Greeks, ancient man had little interest in the structure of materials. He accepted a solid to be just that—a continuous uninterrupted substance. Some of the Greeks thought that if a person began to subdivide a piece of material such as copper, he could do so indefinitely. It was among these people that the idea of continuous matter was fostered. Others reasoned that there must be a limit to the number of subdivisions that one could make and still retain the original characteristics of the material being subdivided. They held fast to the idea that there must be a basic particle upon which all substances are built. Both of these arguments were equally valid at that time, for there was still no means available to determine which faction was correct. Mankind did not know the answer to this question until the nineteenth century.

It was not until 1805 that John Dalton proposed his theories concerning the nature and behavior of matter. He proposed that all matter is composed of invisible, solid, indestructible particles.

Composition of Matter

It was near the middle of the seventeenth century that Robert Boyle phrased the first definition of an elemental substance. He stated that an element is a substance that cannot be decomposed into simpler substances. There are over 100 known elements with the possibility of the discovery of many more. They range from the abundant elements such as silicon, carbon, and oxygen to the rare elements such as lanthanum, samarium, and thulium which are extremely difficult to process. During World War II many elements were synthesized (man-made). The names of the man-made elements are interesting because in many cases they indicate their origin by their names. Elements such as americium, californium, and berkelium are examples of elements of this type.

To make the discussion of elements and the subsequent material more meaningful, a list of elements called the periodic table is provided in the Appendix. Notice that this table is separated into vertical and horizontal columns that form blocks into which are placed symbols that represent the different elements. The vertical columns are called groups, and the horizontal rows are called periods. The symbol used for iron is Fe. It is located in period 4, group VII. The symbol used to represent copper (Cu) is also in period 4, but it is in group IB. Notice that the elements in the *B* groups and group VII are all heavy metals, the elements in groups IA and IIA are light metals, and the elements in groups IIIA and VIIA are nonmetals. The elements in the far-right group are called the inert gases. They are called inert because they will not combine chemically with other substances. Laser action has been observed in almost all of the inert, or noble, gases. The elements boron, silicon, germanium, arsenic, antimony, tellurium, and polonium are called metalloids because under certain conditions they can possess the properties of metals and nonmetals. The two remaining columns contain the lanthanum series, which are the rare earth elements, and the actinium series, part of which include the man-made elements 95 through 102.

Although many substances are composed of a single element, a far greater number of substances are composed of a combination of different elements. When two or more different elements are chemically combined, they form *compounds*. A common example of a compound would be a substance such as ammonia, which is composed of the element hydrogen and the element nitrogen. The process whereby the elements are chemically combined to form a compound is called *synthesis*. During the synthesis of ammonia, one part of the nitrogen element is combined with three parts of the hydrogen element. A compound once formed by synthesis can also be examined and broken down into its elements. This process of examination and reduction is known as *analysis*.

As elements such as hydrogen and nitrogen are chemically combined to form a compound, they lose their individual identity. A most vivid realization of this fact can be noted when visualizing white crystalline sugar. This compound consists of the black, solid element carbon, and two colorless gaseous elements, oxygen and hydrogen. Thousands of compounds are known, each of which possesses definite chemical and physical properties that enable it to be distinguished from other compounds. The almost limitless combinations of elements to form compounds has led to discoveries of the many substances which have become a part of our daily lives. A few common examples of compounds are salt, wood, and limestone. Examples of laser compounds are ammonia, a gas; carbon dioxide, a gas; and gallium arsenide, a solid.

When various elements or compounds are mechanically combined without the occurrence of a chemical change, the result is called a *mixture*. The component elements or compounds of a mixture do not lose their chemical or physical properties. Though the mixture can acquire an appearance that differs from any of its parts, each ingredient that blends into forming the mixture will retain its identity. Thus, it is possible to easily separate a mixture into its individual parts. Gas lasers often use gas mixtures, such as helium-neon or argon-oxygen.

There are two types of mixtures—heterogeneous and homogeneous. A heterogeneous mixture, such as may be formed by combining sand and gravel, is one which does not have a uniform blending of ingredients. Homogeneous mixtures are those that have a uniform composition such as homogenized milk or sugar dissolved in water. Homogeneous mixtures are sometimes called solutions. Gas mixtures are also homogenous mixtures.

The discovery of the many substances that have become a part of our lives would not have been possible without a great deal of study of the elements. Since the elements are the fundamental substance of all matter, the development of any new product must be based on a knowledge of these substances. The elements cannot be decomposed into a simpler substance; therefore, the dissimilarity between them can only be explained by assuming each element to consist of basic particles. This basic particle is called an atom. While the atoms of a given element are similar, the atoms of different elements will have different characteristics.

The Atom

An atom is defined as the smallest particle of an element that retains all of the properties of the element. The following is Dalton's conception of the atom:

1. All materials are composed of minute indestructible particles called atoms.
2. The atom is the smallest component part of an element that enters into a chemical reaction.
3. All atoms of a given element are exactly the same in weight, shape, and size.

The atom is the smallest part of an element that enters into a chemical change, but it does so in the form of a charged particle. These charged particles are called ions, and they are of two types—positive and negative. One of the properties of charged ions is that ions of the same charge tend to repel one another, whereas ions of unlike charge will attract one another. The term charge has been used loosely. At present, charge will be taken to mean a quantity of electricity which can be one of two kinds, positive or negative. Gas lasers often employ ions.

Molecule

The combination of two or more atoms to form the smallest part of a compound comprises a structure known as a molecule. For example, when the compound water is formed, two atoms of hydrogen and one atom of oxygen combine to form a molecule of water. A single molecule is very small and is not visible to the naked eye. Therefore, a few drops of water can contain as many as a million molecules. A single molecule is the smallest particle into which the compound can be broken down and still be the same substance. Once the last molecule of a compound is divided into atoms, the substance no longer exists. The molecular nature of laser action can be inferred from the fact that the term *maser* is sometimes taken to mean *molecular* amplification by stimulated emission of radiation, instead of *microwave* amplification by stimulated emission of radiation. The laser can then be referred to as an *optical maser*. Sometimes the action of lasers and masers is referred to as *aser* action in an attempt to cover all the bases. The terms maser and laser, however, are now well entrenched and are the ones used throughout this book.

An element, while being composed of like atoms, is also considered to have a molecular structure. The term molecule, when applied to an element, designates the minutest portion of an element that can exist under normal conditions and still retain the characteristics of the element. A molecule of an element most often contains only one atom, and is called a monatomic molecule. Examples of monatomic molecules are molecules of iron, gold, and copper. There are some elements, however, whose molecules cannot exist under normal conditions as monatomic and must consist of a combination of atoms. Molecules having two atoms are called diatomic molecules. Examples of diatomic molecules are oxygen, hydrogen, and nitrogen. The molecules of some elements called polyatomic molecules are composed of many atoms. As an example, a molecule of sulphur contains as many as eight atoms.

All atoms of all elements are similar because their contents are alike. Atoms are composed of minute particles, the discovery and the characteristics of which will now be discussed. Atoms basically consists of electrons, protons, and neutrons.

An Atomic Model

Once the basic constituents of the atom are known, an attempt can be made to construct a suitable atomic model. This model must accurately represent and be compatible with all of the facts known at the time the model is constructed. Dalton viewed the atom as a small indestructible sphere having the ability to become firmly attached to other atomic spheres. Later and more advanced experimentation proved that tiny charged particles could be removed from inside the atom. As a result, Dalton's model could no longer be considered satisfactory.

Thompson advanced the theory that atoms must have a structure since a fundamental particle can be extracted from them. He envisioned the atom as being a sphere in which were contained a sufficient number of positive and negative charges to make the overall charge of the atom neutral. Thompson's idea that the positive and negative charges were evenly distributed throughout a sphere was disproved in an experiment conducted by Sir Ernest Rutherford.

In this experiment, a narrow bean of alpha particles (positive double-charged helium ions) was obtained from a sample of radium and directed through a small hole in a lead block toward a thin sheet of gold foil. If the atom were constructed as Thompson visualized, the positive alpha particles should have had their paths deflected by small

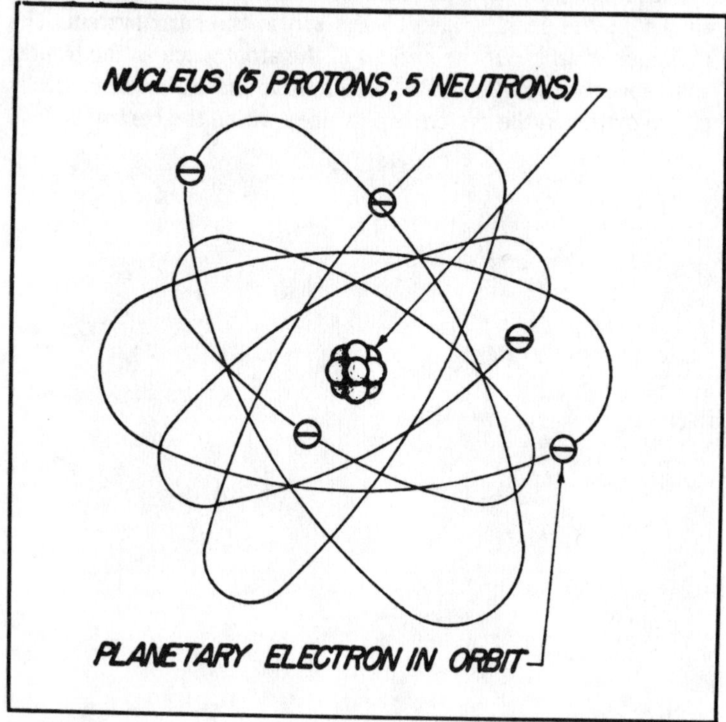

Fig. 2-3. Boron atom.

amounts due to the positive charge distributed evenly through the atom. The results were hardly what was expected. Rutherford found that most of the alpha particles went right through the gold foil without being deflected at all. The remaining particles received large amounts of deflection, some as high as 180°. This could only be explained by assuming that all of the positive charge in the atom was concentrated in one area away from the negative charge. Any alpha particle coming close to this center of charge would be severely deflected, while one passing some distance away would go through the foil undeflected.

From the results of Rutherford's experiments, emerged our present concept of the structure of the atom. The atom is now believed to consist of a group of positive and neutral particles (protons and neutrons) called the nucleus, surrounded by one or more negative orbital electrons. Figure 2-3 shows the arrangement of these particles for an atom of the element boron. This concept of the atom can be likened to our solar system in which the sun is the massive central body, and the planets revolve in orbits at discrete distances from the sun. The nucleus commands a position in the atom similar to the position held by the sun in the solar system. The electrons whirl about the nucleus of the atom much as the planets whirl about the sun. In both the solar system and the atom practically all the matter in the system is contained within the central body.

Electricity, Magnetism and Waves

The laser is a new type of electromagnetic wave generator, but it uses the same basic forces of magnetism and electricity as other generators. It is important to know about these forces and how they interact to produce electromagnetic waves. We also need to learn more about the electromagnetic spectrum, the place of maser and laser energy in the spectrum, the kinds of equipment used in different parts of the spectrum, and why laser and maser equipment are so useful for communications and other purposes.

To understand how electrons emit energy in lasers and masers, we need to know about the two kinds of energy possessed by electrons: kinetic energy and potential energy. We begin this chapter with a consideration of the nature of these types of energy and how energy relates to charged particles such as electrons.

WORK AND ENERGY

In everyday life, work is commonly thought of as anything that requires physical or mental exertion. In the field of physical science, however, work must be defined more precisely: *work is the product of displacement and force*. The amount of work accomplished by movement of an object can be calculated by the equation

$$W = Fd$$

Where W is the work performed, F is the force applied in direction of displacement, and d is the distance through which the force acts.

It is important to notice that according to the precise definition of work, no work is accomplished unless the force applied causes a change in position of a stationary object, or a change in velocity of a moving object. A man can tire himself by pushing against a heavy wooden box, but unless the box moves as a result of his efforts, he has accomplished no work.

Kinetic Energy

Whenever work is accomplished on an object, energy is consumed (changed from one kind to another). If no energy is available, no work can be performed. Thus, energy is the ability to do work.

One form of energy is that which is contained by an object in motion. In driving a nail into a block of wood, a hammer is set in motion in the direction of the nail. As the hammer strikes the nail, the energy of motion of the hammer is converted into work as the nail is driven into the wood. Energy contained by an object due to its motion is called ***kinetic energy***.

Potential Energy

In addition to kinetic energy, an object can contain energy by virtue of its position within a system. A hammer and the earth form a system of masses in which exchanges of energy can take place through the medium of the earth's gravitational field. Assume that the hammer is suspended by a string in a position meter above a nail. As a result of gravitational attraction, the hammer will experience a force pulling it downward toward the center of the earth. If the string is suddenly cut, the force of gravity will pull the hammer downward against the nail, driving it into the wood. While the hammer is suspended above the nail it has the ability to do work because of its elevated position in the earth's gravitational field. Since energy is the ability to do work, the hammer contains energy.

Energy contained by an object due to its position is called ***potential energy***. The amount of potential energy available is equal to the product of the force required to elevate the object and the height to which it is elevated.

Potential Difference

In most cases the total potential energy in a system is of little practical importance. A good example of this is the old-fashioned cuckoo clock. Figure 3-1 shows such a clock in which the gear mechanism used to turn the hands is operated by a slowly falling

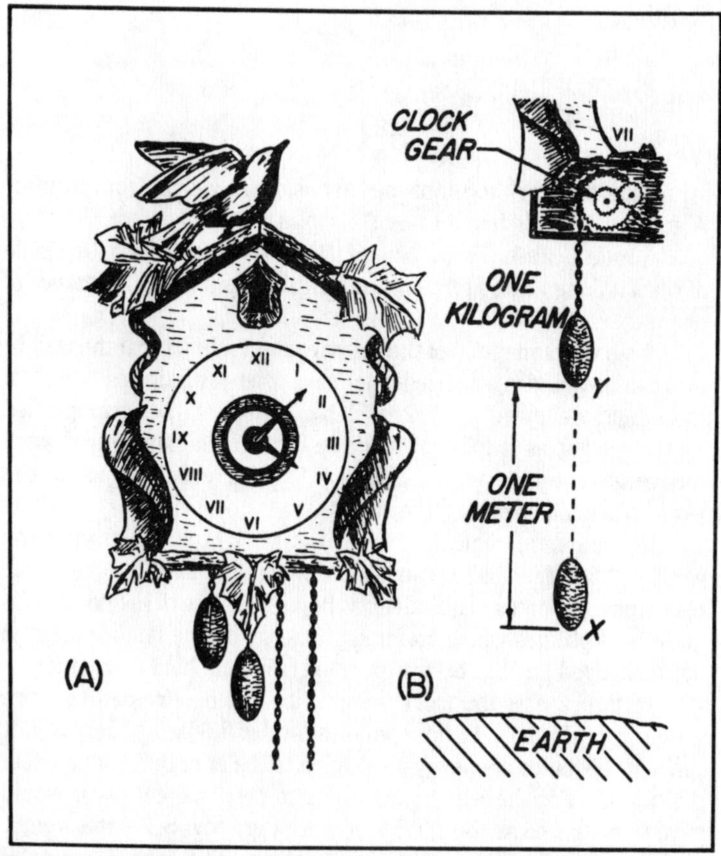

Fig. 3-1. Weight-driven cuckoo clock.

weight. In Fig. 3-1 a metal weight is suspended on a chain beneath the clock, so that the gear will be rotated as the weight pulls downward on the chain. With a given length of chain, the weight can only fall a short distance to position X, its lowest possible position. To begin, assume that the weight, having a mass of one kilogram, is at position X. If the weight is then raised against the force of gravity from position X to position Y, a distance of one meter, work will have been performed on the weight. The force required to overcome gravity is known to be 9.8 newtons for a mass of one kilogram. The work accomplished can be calculated as follows:

Given:

$$F = 9.8 \text{ newtons and } d = 1.0 \text{ meter}$$

Solution:

$$W = fd$$
$$W = 9.8\ \text{N} \times 1.0\ \text{m}$$
$$W = 9.8\ \text{J}$$

Thus, the work accomplished in raising the weight from position X to position Y is found to be 9.8 joules.

A joule is a unit of work named after a British physicist. One joule of work is done when a force of one newton acts through a distance of one meter.

A close examination of the above example shows that the weight now has capabilities which it did not have when it was in position X. If the weight is allowed to fall from Y to X, work will be accomplished by the weight as it falls. Neglecting friction, the amount of work recovered when the weight falls is 9.8 joules, exactly equal to the work expended in raising the weight.

It is no mere coincidence that the work required to raise the weight (9.8 joules) and the amount of work the raised weight is able to perform are equal. This equality stems from one of the fundamental laws of physics which states that energy can be neither created nor destroyed but can be transformed from one kind to another.

In the case of the clock weight, it must be stressed that the weight in position Y did not contain a total of 9.8 joules of energy. It *gained* 9.8 joules of energy in addition to the energy it contained in position X. For the weight to have zero potential energy it would have to be placed at the earth's center of gravity. Since the weight could not possibly fall all the way to the earth's center, the total potential energy of the weight could never be fully utilized. Of far greater interest is the additional energy that is added to the system as the weight is raised from its lowest position to its highest position since this represents the amount of energy which can be recovered as work.

To simplify problems dealing with potential energy, the lowest position of a body is used as a reference point and the body is considered to have zero potential energy when in this position. (This simplification is similar to the system used in measuring altitude where sea level, rather than the center of the earth, is used as the zero reference.) The potential energy of the upper position of the body could then be computed with respect to the lower position. The result would be the difference of potential between the two positions and a true measure of the work that may be recovered. In future calculations the difference of potential energy will be considered rather than the absolute potential energies.

The concept of energy ranks as one of the greatest inventions of all time. It is central to the mechanism of lasers and masers, as later chapters show.

The Electric Field

From the study of electrostatics it was learned that a field of force exists in the space surrounding a quantity of charge. This field of force, although electrical in nature, is very similar to the earth's gravitational field. As a result of this similarity, many of the laws developed for the gravitational field apply equally as well to the electric field.

For the purpose of explanation assume that a positive charge of magnitude Q_1 exists at an isolated point in space as shown in Fig. 3-2.

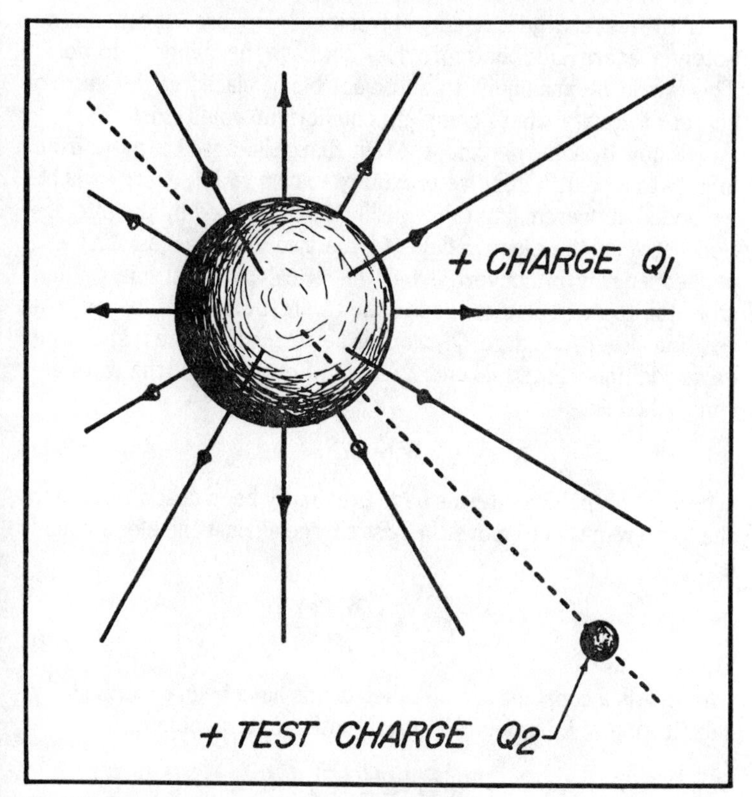

Fig. 3-2. Electric field surrounding a positive charge.

As illustrated in the drawing, the space around charge Q_1 is filled with an electric field of force. The field of force is theorized to consist of electrostatic lines of force, each line of which starts at the surface of the charge and extends outward to infinity. This field is seen to be very dense in close proximity to charge Q_1, but diminishes rapidly in intensity as the distance from Q_1, is increased.

Electrical Potential Energy

If a positive test charge of magnitude Q_2 is inserted into the electrical field of charge Q_1 (described above), a situation exists similar to that of the weight placed into the gravitational field of earth. If charge Q_2 is initially placed into the field at an infinite distance from charge Q_1, little or no force would be exerted on Q_2 by charge Q_1. For practical purposes the intensity of the field about charge Q_1 could be considered to be zero at this remote distance. Since the test charge is resting at a point of zero force, it contains no potential energy and therefore does not have the ability to do work. This would be analogous to an object being placed at the earth's center of gravity where complete equilibrium would exist.

If now by some method, the test charge is caused to move from infinity to a point of relative proximity to charge Q_1, energy will be expended in overcoming the repelling effect caused by the interaction between the electric fields of the two like charges. This expended energy is converted into electrical potential energy and stored in the test charge. Each time the test charge is moved to a position closer to charge Q_1, work is performed and the test charge gains additional potential energy. From Equation 3-1, the work accomplished is:

$$W = Fd \tag{3-1}$$

where all of the given terms have previously been described. Since the force required to move the test charge against the electric field is:

$$F = k\frac{(Q_1 Q_2)}{d^2} \tag{3-2}$$

where k is a constant and all other terms have been described. By substitution of Equation 3-2 into Equation 3-1 we obtain:

$$W = k\frac{(Q_1 Q_2)\ d}{d^2}$$

therefore,

$$W = k\frac{(Q_1 Q_2)}{d} \tag{3-3}$$

Electrical Potential Difference

Refer again to the charges in the previous example. If the test charge Q_2 is held in a stationary position in the electric force field of charge Q_1, a condition exists which is comparable to the situation of the raised clock weight (Fig. 3-1). The raised weight contained potential energy due to its elevated position in the earth's gravitational field. As the weight was allowed to slowly fall, its potential energy was converted into useful work as it turned the gears and hands of the clock. A similar comparison can be made to the stationary test charge since it also contains potential energy. If the test charge is allowed freedom of movement, it will be repelled from charge Q_1 along the dotted line as shown in Fig. 3-2. Since work was defined as force times distance, the test charge accomplished work as it is repelled from charge Q_1. All of the energy that was expended in moving Q_2 toward Q_1 was stored in Q_2 and will now be recovered (in the form of kinetic energy) when the test charge is allowed to be repelled.

Gravitational potential energy is gained when an object is moved against the earth's gravitational field. In Fig. 3-3A the surface of the earth could arbitrarily be allowed to represent zero potential. If the force F were to cause the one-kilogram weight to be raised a distance of one meter, work would be performed on the weight and it would now contain gravitational potential. Since a one-kilogram mass exerts a force of 9.8 newtons as a result of gravity, this gravitational potential may be calculated as follows:

$$W = Fd$$
$$W = (1\text{ kg})\,(9.8\text{ N/kg})\,(1\text{m})$$
$$W = 9.8\text{J}$$

Gravitational potential energy is expressed as joules per kilogram (for which no unit has been devised), therefore, the weight in Fig. 3-3A has gained 9.8 joules of potential energy with respect to ground (earth).

A similar situation also exists for the electrical system represented in Fig. 3-3B. If a quantity of charge is to be moved from terminal T_1 to terminal T_2 against an electrical force, work must be done on the charge, and the charge will gain potential energy. In an electrical system, the electric potential is expressed as joules per

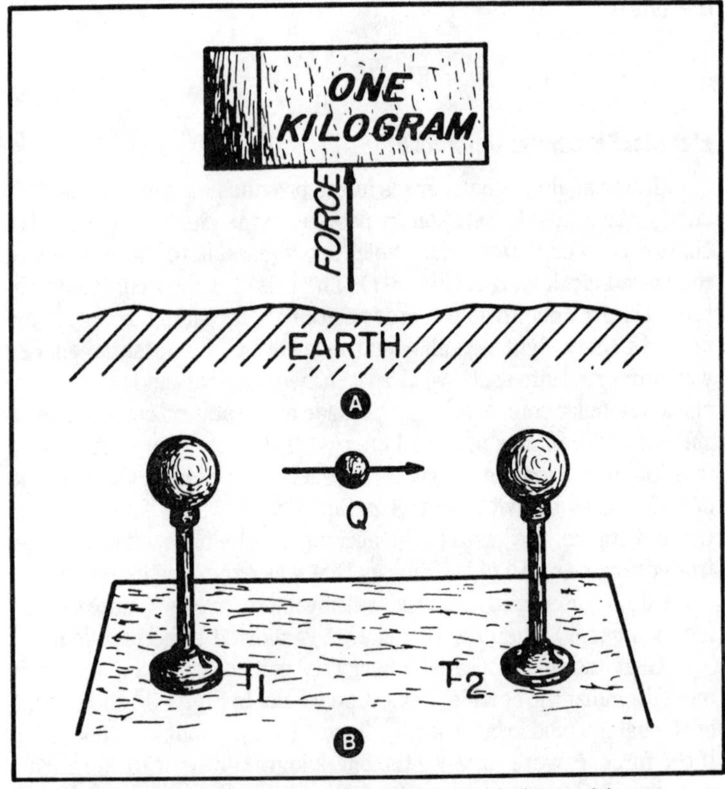

Fig. 3-3. Comparison between gravitational and electrical potential.

coulomb and represents work (joules) per unit charge (coulombs). A potential of one joule per coulomb is called one volt in honor of Alessandro Volta, who discovered one of the first practical methods of generating a continuous electrical potential. The symbol used for voltage in equations is the letter *E*. (The letter *V* is the abbreviation used for volt as a unit of measure.) If the joules of work (*W*) and the coulombs of charge (*Q*) are known, potential E in volts can be calculated by the following formula:

$$E = \frac{\text{joules}}{\text{coulombs}}$$

$$E = \frac{W}{Q}$$

In Fig. 3-3B, if three coulombs of charge are moved from T_1 to T_2, and 180 joules of work are expended, the voltage at T_2 would be:

$$E = \frac{W}{Q} \qquad E = \frac{180}{3}$$

$$E = 60 \text{ volts}$$

This indicates that the potential at T_2 differs from the potential of T_1 by 60 volts, or that there is an electrical potential difference between T_2 and T_1 of 60 volts.

Polarity of Potential

In the discussion above, nothing was said about the polarity of the potential developed by the movement of charges from T_1 to T_2. Assuming that the charges were positive, T_2 would become 60 volts positive with respect to T_1. In many sources of potential the charges to be moved are electrons. If this were the case in Fig. 3-3B, the charge deposited on terminal T_2 would be negative. As a result, T_2 would be 60 volts negative with respect to T_1.

Symbols and Terms

In the previous sections the concept of a difference of potential was developed. In most electronic circuits only the difference of potential between two points is important, and the absolute potentials of the points are of little concern. Very often it is convenient to use one standard reference for all of the various potentials throughout a piece of equipment. For this reason, the potentials at various points in a circuit are generally measured with respect to the metal chassis on which all parts of the circuit are mounted. The chassis is considered to be at zero potential and all other potentials are either positive or negative with respect to the chassis. When used as the reference point the chassis is said to be at ground potential. Ground reference, abbreviated GND, is usually represented by one of the symbols shown in Fig. 3-4.

Occasionally rather large values of voltage can be encountered, in which case the volt becomes too small a unit for convenience. In a situation of this nature, the kilovolt (kV), meaning 1000 volts, is frequently used. For example, 20,000 volts would be written as 20 kV. In other cases the volt may be too large a unit, as when dealing with very small voltages. For this purpose the millivolt (mV), meaning one-thousandth of a volt, and the microvolt (μV), meaning one-millionth of a volt are used. For example, 0.001 volt would be written as 1 mV, and 0.000025 volt would be written as 25 μV.

In the everyday language of electronics the number of volts between two points is expressed in several different ways, some of

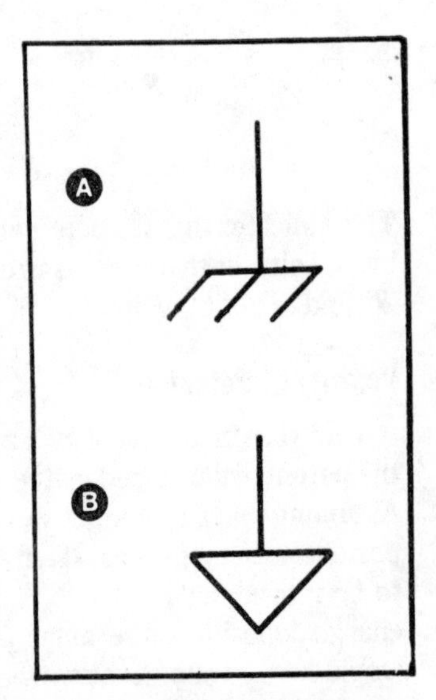

Fig. 3-4. Symbols for ground.

which are: voltage, potential, potential difference, and electromotive force (emf). Strictly speaking, each of these terms indicates a specific quantity; however, they are quite frequently used interchangeably. Emf for example, should only be used when referring to the force which causes charges to move through a source of voltage.

MAGNETIC FIELDS AND EMF

It has been shown that a concentration of charge develops a difference of potential (measured in volts) between it and some reference. Under certain conditions, this difference of potential is capable of accomplishing work. Now we examine the various methods by which a difference of potential or voltage may be generated.

A charge could be produced on a rubber rod by rubbing the rod with cat fur. Due to the friction involved, the rubber rod assumes a negative potential and the fur becomes positive. These quantities of opposite charge constitute a difference of potential between the rod and the fur. The electrons comprising this difference of potential are capable of doing work if a discharge is permitted to occur.

To be a practical source of voltage, the potential difference must not be allowed to dissipate, but must be continuously maintained. As one electron leaves the concentration of negative charge, another must be immediately provided to take its place or the charge will eventually diminish to the point where no further work can be accomplished. A voltage source, therefore, is a device which is capable of supplying and maintaining voltage while some type of electrical apparatus is connected to its terminals. The internal action of the source is such that electrons are continuously removed from one terminal, which becomes positive, and simultaneously supplied to the second terminal, which assumes a negative charge.

Electromotive force is that force which moves charges within a voltage source. This name is misleading, however, because electromotive force cannot be measured in the conventional force units of newtons or pounds but is measured in volts. Thus, potential difference and electromotive force (emf) are both measured in volts even though they are slightly different quantities and can have different values in a given source.

Generation by Magnetism

One of the most widely used methods of generating an emf is that of generation by electromagnetic induction. In this process an electromotive force is caused to exist in a conductor by the interaction of the conductor with a magnetic field of force. Figure 3-5 shows a length of copper conductor as it is passed through the poles of a horseshoe magnet. As the conductor passes between the poles of the magnet, it must travel through the field of force between the poles. This field is represented by lines of force shown leaving the north pole and entering the south pole of the magnet. Thus, as the conductor moves through the field it cuts the lines of force.

A conductor has many free electrons. These electrons are not firmly attached to any particular atom, but are free to wander about within the conductor. If the conductor is made to move, it carries these free electrons along with it, causing the electrons to move at the same speed and in the same direction as the conductor. Thus, as the conductor in Fig. 3-5 moves through the magnetic field, the free electron which it contains moves with it.

Whenever an electron moves, it generates a magnetic field whose lines of force appear as concentric circles about the electron as illustrated in Fig. 3-6. To determine the direction of the field encircling the electron, visualize the left hand placed around the electron as shown in Fig. 3-6, with the thumb pointing in the direction in which the electron is traveling. The curled fingers then point in the direction of the magnetic field about the electron (see Fig. 3-6). The direction of this field is shown by the arrowheads on the lines of force encircling the electron.

In Fig. 3-7 the lines of force about the electron are shown as the electron enters the field between the poles of a magnet. As shown in

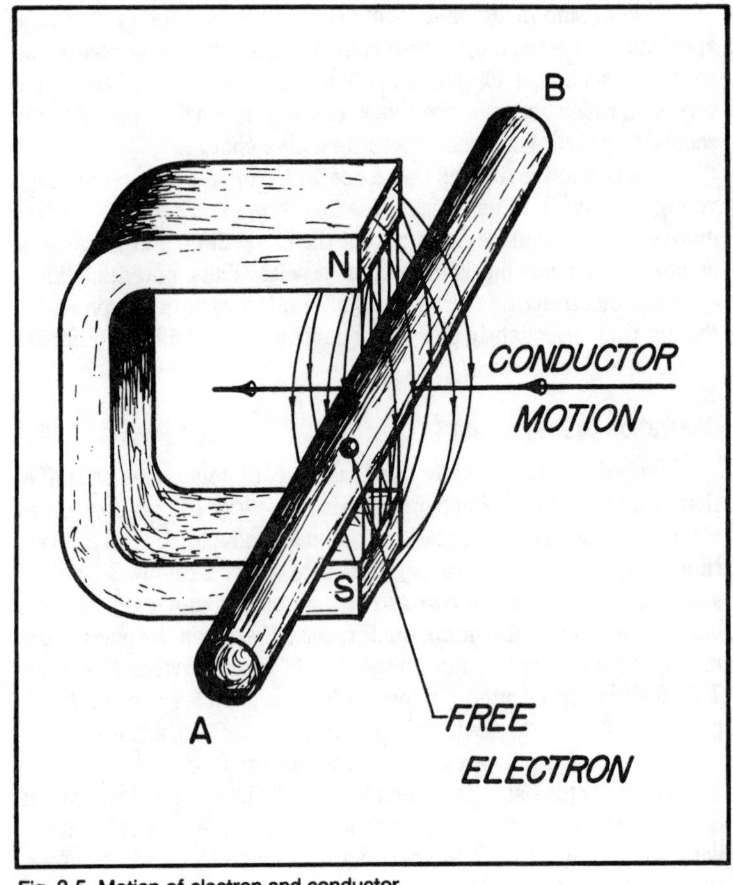

Fig. 3-5. Motion of electron and conductor.

the diagram, the lines about the electron will be traveling in the same direction on the left-hand side of the electron, and in opposite directions on the right-hand side of the electron.

Lines of force traveling in the same direction repel one another, and lines traveling in opposite directions attract. This being true, the two sets of lines on the left-hand side of the electron (Fig. 3-7) will repel each other while those on the right will attract. This action will produce a force on the electron directed to the right of the diagram.

Since the electron is free to move within the conductor, it will be forced to the end of the conductor labeled *B* in Fig. 3-8. Because all electrons are similar, all free electrons in the conductor will have a tendency to move from *A* to *B* as the conductor is moved between the poles of the magnet.

The net result of the movement of the conductor into the field of the magnet is the forced or directed migration of millions of electrons through the conductor from *A* to *B*, as illustrated in Fig. 3-8. The electrons, therefore, are repelled from the *A* end of the conductor and will accumulate at the *B* end, creating a difference of potential between these two points. This process of generating a voltage is called ***electromagnetic induction***, and the voltage generated is called an ***induced voltage***. The majority of the world's commercial electric power is produced by this method, and some early radio transmitters were actually high-frequency ac generators.

Quanta and Light

In 1905, Albert Einstein developed the theory that a beam of light consists of small bundles of energy called light quanta or photons. This theory was so radical compared to other contemporary theories concerning light energy that it was not generally accepted until 1916 when additional experiments by an American physicist,

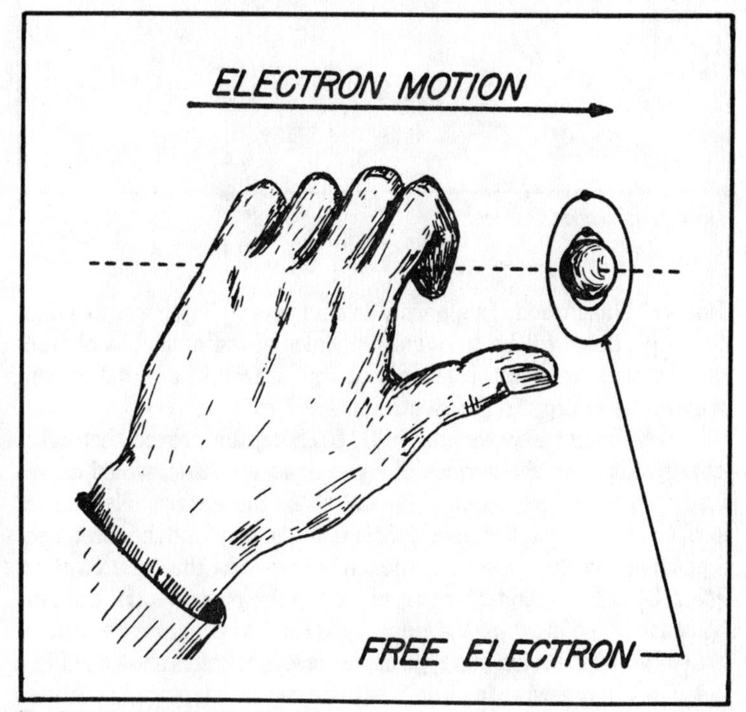

Fig. 3-6. Field about moving electron.

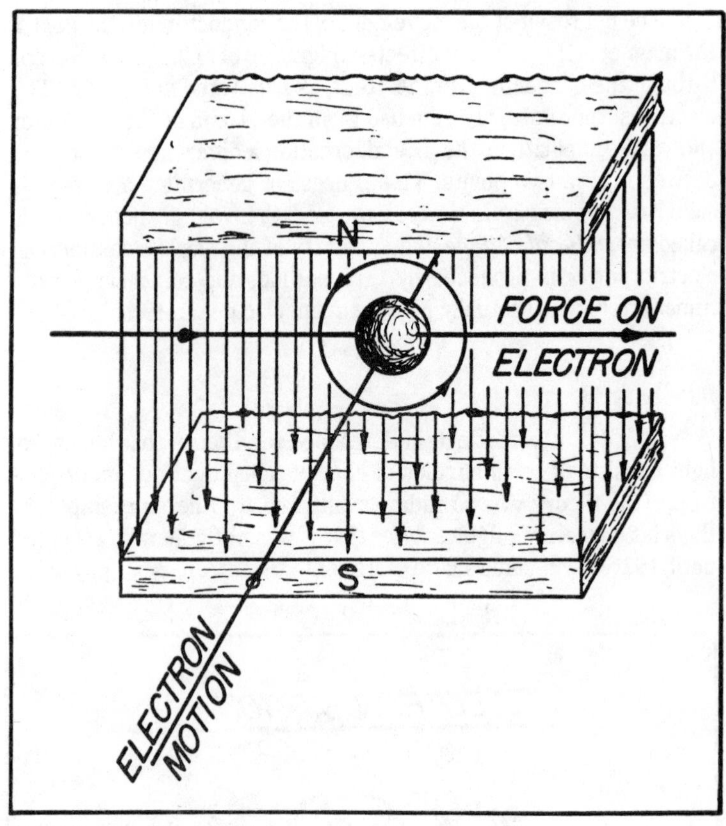

Fig. 3-7. Interaction of fields.

Robert Millikan, added support. The energy of a photon was found to be proportional to the frequency or color of the light. If a photon collides with an electron at or near the surface of a metal, it can transfer its energy to the electron.

Subsequent experiments by J. J. Thompson proved that light energy falling on the surface of a metal, such as zinc, would cause electrons to be forced out of the metal. As the electrons leave the surface of the zinc it becomes deficient in electrons and thus assumes a positive charge. This phenomenon is known as the *photoelectric effect*. By properly combining layers of metal, a photoelectric cell can be constructed in which light energy is used to generate an emf.

The construction of a typical photoelectric cell is shown in Fig. 3-9. The cell consists of an iron base plate on which is placed a coating of selenium. Next a layer of material is placed over the selenium

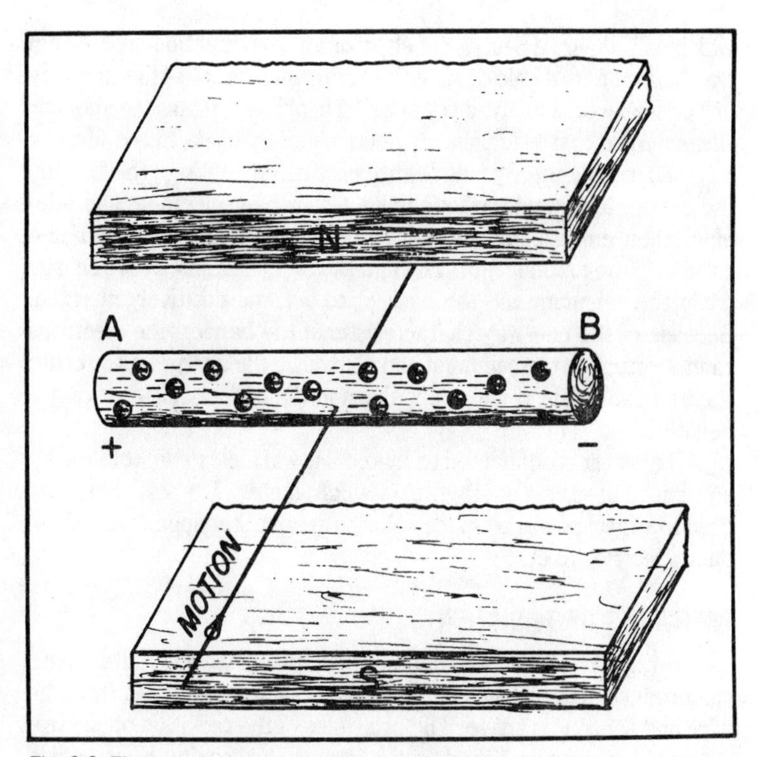

Fig. 3-8. Electron displacement within a conductor.

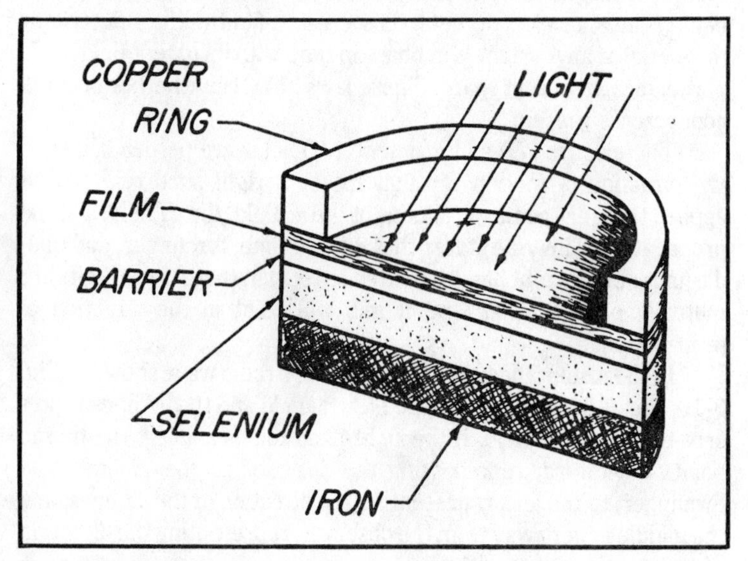

Fig. 3-9. Photoelectric cell.

which will allow the passage of electrons in one direction only. A thin transparent film of gold or silver is then formed on top of this one-way or barrier layer. Finally, a copper ring is pressed against the gold or silver film so that electrical connection can be made to the film.

When light energy falls on the cell it travels through the film and the barrier layer and strikes the light-sensitive selenium. The selenium then emits electrons which travel through the barrier and collect on the metallic film. The film thus acquires a negative charge while the selenium and the back plate become positively charged. Because of the one-way characteristic of the barrier, the electrons cannot return to the selenium and neutralize the charge. As a result, the iron back plate remains positive and the copper collector ring is negative.

If a meter is connected to this cell, it will indicate light intensity and can be used as a photographer's light meter. The solar batteries used to supply power to space vehicles are another application of the photoelectric effect.

PRINCIPLES OF RADIATION

A radio-frequency (rf) current flowing in a wire of finite length can produce electromagnetic fields that can be disengaged from the wire and set free in space. The principles of the radiation of electromagnetic energy are based on the laws that a moving electric field creates a magnetic field, and conversely, that a moving magnetic field creates an electric field. The created field (either electric or magnetic) at any instant is in phase in time with its parent field but is perpendicular to it in space. These laws hold true whether or not a conductor is present.

The electric (E) and magnetic (H) fields are perpendicular to the direction of motion through space. A right-hand rule can be applied that relates the directions of the E field, the H field, and the propagation. This rule states that if the thumb, forefinger, and middle finger of the right hand are extended so that these three digits are mutually perpendicular, the thumb will point in the direction of propagation.

In the instantaneous cross-section of a radio wave shown in Fig. 3-10, the E lines represent the electric field and the H lines represent the magnetic field. If the right-hand rule is applied, the thumb points downward, representing the direction of the E lines; the forefinger, to the left, representing the direction of the H lines; and the middle finger away from the observer, representing the direction of propagation.

When rf current flows through a transmitting antenna, radio waves are radiated from the antenna in all directions in much the same way that waves travel on the surface of a pond into which a rock has been thrown. It has been found that these radio waves travel at a speed of approximately 186,000 miles per second (300 million meters per second). The frequency of the radio wave radiated by the antenna will be equal to the frequency of the rf current.

Because the velocity of the radio wave is constant regardless of its frequency, to find the wavelength (which is the distance traveled by the radio wave in the time required for one cycle) it is necessary only to divide the velocity by the frequency of the wave. That is,

$$\lambda = \frac{300,000,000}{f}$$

where λ (the Greek letter lambda, used to symbolize wavelength) is the distance in meters from the crest of one wave to the crest of the next, f the frequency in hertz (cycles per second), and 300,000,000 the velocity of the radio wave in meters per second. This relationship is important in radio communications. It can also be expressed as

$$f = \frac{300}{\lambda}$$

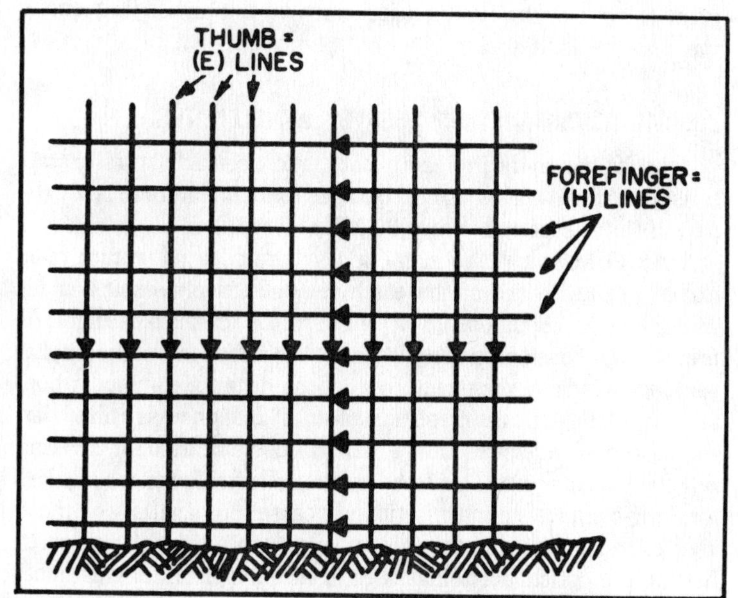

Fig. 3-10. Instantaneous cross section of a radio wave.

where f is in megahertz, λ is in meters, and 300 is the velocity of propagation of the radio wave in millions of meters per second.

For example, the frequency of the current in a transmitting antenna that is radiating an electromagnetic wave having a wavelength of 2 meters is 300/2, or 150 megahertz. Radio waves are usually referred to in terms of their frequency. Maser outputs are referred to in terms of their frequency or their wavelength. Laser outputs are referred to in terms of their wavelength.

UPWARDS TOWARD MASERS AND LASERS

The rapid expansion of radio broadcasting and communications created the serious problem of interference between stations. To alleviate this situation, the governments of various countries have held many conferences since 1903, when they first reached an international agreement relating to the assignment of radio-frequency channels for various radio services. The field of radio increased rapidly and by the early 1930s the radio channels between the lowest practical frequency (about 15 kHz) and the highest frequencies that had proved useful for long-distance communication were so congested as to hamper their usefulness. Thus the radio industry, radio amateurs, and various government and military agencies were stimulated to explore the region above 30 MHz. In the early 1920s a great amount of the experimental work on the higher frequencies was done by amateurs.

SPECIAL ADVANTAGES OF SHORTER WAVELENGTHS

In addition to the problem of congestion, certain natural characteristics of radio waves also encouraged experimenters to study the part of the rf spectrum above 30 MHz.

At 30 MHz and above, the ionosphere does not return radio waves to the surface of the earth very effectively, except under rather unusual conditions. In fact these waves act much like light. At first this was considered a serious limitation since most of the emphasis in early radio development was on long-distance communication, far beyond the optical horizon (line of sight). It soon was realized that the shorter wavelengths, above 30 MHz, could be used for covering relatively local areas. This freed some additional lower frequencies for long-distance communication. Because propagation of these shorter radio waves did not reach points on the surface of the earth beyond the optical horizon as seen from the transmitting antenna, stations could operate on the same assigned frequency without interference, if they were separated far enough apart geographically. This principle already was being used on the broadcast band and other radio services operating in the region above 30 MHz.

A second effect of the decrease in wavelength as the frequency is increased is connected with the phenomenon of radiowave reflection. All electromagnetic waves, such as radio, light, and heat, can be reflected, but how well they are reflected depends on a number of different factors. One factor is the relationship between the length of the wave considered and the physical size of the reflecting object. In general, an object must be a reasonable fraction (1/10 to 1/5) of a wavelength long in one dimension to reflect radio waves effectively. Objects of one or more electrical half-wavelengths reflect best if other factors are equal. Therefore, the shorter the wavelength, the smaller the object that can reflect the waves effectively. For example, a wave ten meters long will be reflected readily from objects that would have little effect on a wave 100 meters long. Similarly, waves one meter long are reflected readily from objects the size of a car or an airplane. This is one of the basic principles on which radar operates, and its importance hardly can be exaggerated. Since many of the objects it is desired to detect with radar are under five to ten meters in length, radio waves five meters or less in length will be more effective in detecting them than longer waves. Actually, most radar equipment uses wavelengths much shorter than five meters, ranging down to one centimeter or less. Earlier equipment used wavelengths up to ten meters, and much of the radar used in World War II operated at about 500 MHz, a wavelength of 0.6 meter.

Another factor directly related to the wavelength is the physical size of the equipment used to generate the rf energy, and the antenna needed to radiate it effectively. Both of these factors can be made smaller in direct proportion as the wavelength is made shorter. For example, a half-wave antenna for a station operating in the broadcast band requires a steel tower hundreds of feet high, but at 500 MHz an aluminum rod 30 centimeters long is sufficient. Obviously, equipment for the shorter wavelengths can be made more compact and portable because of this relationship between physical size and wavelength. This relationship between wavelength and physical size makes possible the construction of relatively small antennas capable of concentrating the radiated waves into a sharp, narrow beam. The beam, which can be pointed in a desired direction by properly positioning the antenna, aids communication in the desired direction and reduces interference to stations in other directions. Masers and lasers carry this directivity principle to an extreme. This fact and the wide bandwidths of maser and laser systems make them extraordinarily useful for communications.

As we progress through the electromagnetic spectrum to higher and higher frequencies, the components used in producing and handling the frequencies change radically, as Fig. 3-11 shows. As the frequency frontier was expanded into the microwave range, researchers found that conventional lumped-component circuitry—resistors, capacitors, and inductors—dissipated too much of the microwave energy as heat. They were forced to resort to pipes (waveguides) to carry microwaves, special tubes such as klystrons to generate them, and resonant cavities to control them.

As the frequency frontier was pushed still further during World War II, the dimensions of the cavities kept getting even smaller. By the end of the war, the shortest waves that could be generated electronically were about six millimeters (50 GHz). To go much further, the dimensions of the necessary cavity would have to be exceedingly small—they would have to be of molecular size. In fact, it turned out that the molecule could serve as a generator of electromagnetic waves. In 1955, Townes, Gordon, and Zeiger perfected an oscillator in which ammonia gas molecules acted as a multitude of minute radio transmitters at a frequency of 24 GHz. Later developments in the stimulated-radiation technology pushed the frequency many times higher, through the infrared range and into the visible-light range. At last it was possible to electronically generate waves that not only acted like light waves but actually were light waves. The field of electronics had expanded into the field of optics. It was ironic that the field into which electronics had expanded was much older than electronics—but fortunate, because optical instruments

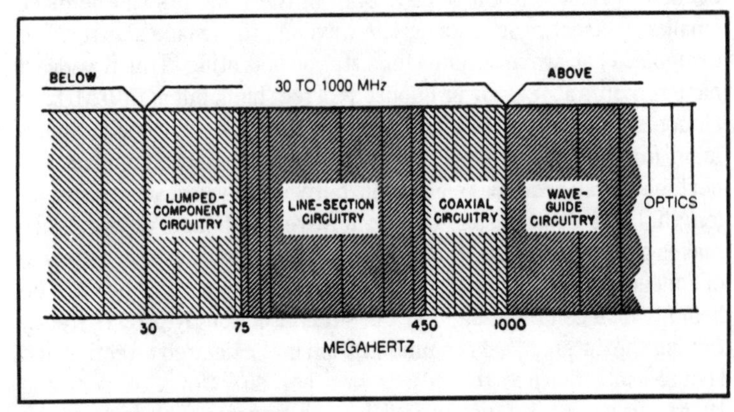

Fig. 3-11. Radio-frequency spectrum and components.

had reached a high state of development, as we shall see in a later chapter.

The ability to electronically generate and amplify light—and a very special kind of light at that—has astounding possibilities. In the years just ahead, quantum electronics will produce quantum leaps in the fields of science, metal fabrication, and communications.

For pure science, the laser principle (based on quantum mechanics, which is covered in detail later) offers a new way of looking at the nature and behavior of matter. The very special light of the laser, which can be made very intense and sharply focused, provides a powerful new tool for spectroscopy—the production and investigation of radiations from objects and materials. Spectroscopy provides a means of analyzing the chemical makeup of a material and studying the structure of atoms and molecules. It is one of the major sources of information about the sun and other celestial objects.

For metal fabricating, the laser offers unprecedented opportunities. The intensity and narrowness of a laser beam make it uniquely suitable for many precise cutting and welding operations on hard, space-age metals.

No one knows where the laser may find its greatest use, but it may turn out to be in communications. A laser beam a quarter of an inch in diameter could replace all existing communications circuits between the East Coast and West Coast of the United States. Several lasers could accommodate the communications needs of the whole world, once the proper modulation and demodulation, and other processing techniques are worked out.

These and other applications of lasers are discussed in later chapters of this book.

4

Light and Radiant Energy

Lasers and masers are basically just producers of light and radiant energy. The light and radiant energy from lasers and masers has certain special properties that make it especially useful. Before these special properties of laser and maser energy can be properly understood, however, you must understand the properties of ordinary light and radiant energy.

When a stone is thrown into a pond, a series of circular waves travels away from the disturbance. These waves radiate in all directions on the surface of the water until they are absorbed or until they are diverted by coming into contact with an obstruction of some sort. Water waves are a succession of crests and troughs; the distance from the crest of one wave to the crest of the next is called the wavelength. A cork floating on the surface of the water bobs up and down with the waves, but it moves very little in the direction of wave travel. Water waves are called *transverse waves* because the motion of the water particles is up and down, or at right angles (transverse) to the direction in which the wave is traveling.

A second form of wave motion, also involving the motion of particles of matter, can be produced by the physical vibration of a body. An example of this type of wave is the sound wave that is produced by striking the tine of a tuning fork. When struck, the tuning fork sets up a vibratory motion. As the tine moves in an outward direction, the air immediately in front of the tine is compressed so that its momentary pressure is raised above that of other points in the surrounding medium. Because air is elastic, this disturbance is transmitted progressively in an outward direction as a compression wave. When the tine returns and moves in the inward direction, the air in front of the tine is rarefied so that its momentary pressure is reduced below that at other points in the surrounding medium. This disturbance is also propagated, but in the form of a rarefaction (expansion) wave, and it follows the compression wave through the medium.

The compression and expansion waves are also called *longitudinal waves*, because the particles of matter that comprise the medium move back and forth longitudinally in the direction of wave travel. Figure 4-1 is a simplified representation of the use of a tuning fork to produce a longitudinal wave. The transverse wave shown below the longitudinal wave is merely a convenient device to indicate the relative density of the particles in the medium, and it does not reflect the movement of the particles. Such a wave is also used to represent the strength of an electromagnetic wave at various points in space; again, it does not reflect the movement of particles.

The third type of wave, the electromagnetic wave, does not involve moving particles of matter at all. Rather, it relies on electric and magnetic force fields. The waves previously discussed cannot be propagated in the absence of a conducting medium—electromagnetic waves are propagated most efficiently in the absence of matter. In the electromagnetic wave, two components are required: an electric field and a magnetic field. These two fields are

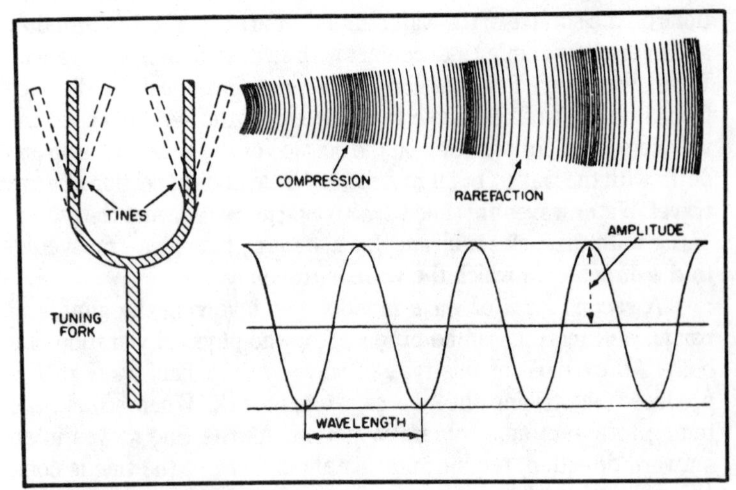

Fig. 4-1. Sound waves.

mutually perpendicular to each other and to the direction of propagation. In the particle waves the velocity of propagation varies with the particular medium and is comparatively slow. In the electromagnetic wave the velocity of propagation is the speed of light (approximately 186,000 miles per second). Examples of electromagnetic waves are heat, light, radio waves, X-rays, cosmic radiation, and ultraviolet rays.

All three types of wave motion obey certain common laws and have certain common characteristics. All are periodic; that is, they all constantly repeat the same pattern so that each succeeding wave is the same as the previous one. Each wave displays the same relationships of wavelength, frequency, period, velocity of propagation, amplitude, and phase. Each wave is subject to the same conditions of reflection and refraction. In the electromagnetic wave, the minima and maxima are correlated with field intensity rather than with particle density or displacement as in the case of the other waves. In the longitudinal wave, the density is related to particle density; in the electromagnetic wave, density is related to the strength of the electric and magnetic fields. In the electromagnetic wave, the electric and magnetic fields are in phase with each other.

A wavelength is the distance measured along the axis of propagation between two points of equal intensity that are in phase on adjacent waves. This length can be represented as the distance between maximum points (maxima) on successive waves.

A line drawn from the source to any point on one of the waves is called a ray, and it indicates the path over which the wave progresses. Although rays do not actually exist, they are frequently used in illustrations as a convenient method of denoting wave propagation. A wavefront is a surface on which all points are in phase. Wavefronts near the source are sharply curved, whereas those at a distance are almost flat; individual rays from a distant source are considered to be parallel.

Modern theory considers light, both visible and invisible, as consisting of quanta (bundles) of energy that move as if guided by waves. The statistical behavior of the quanta depends upon the assumption that the energy of the quanta at any point is, on the average, equal to the intensity of the wave system at that point. Figure 4-2 shows the wavelength in centimeters (cm) of various types of electromagnetic waves including light waves.

The wave theory of light assumes that light is transmitted from luminous (light-emitting) bodies to the eye and other objects by an undulatory or vibrational movement. The velocity of this transmission is approximately 186,300 miles per second, and the vibrations of ether (conducting medium of light in space) are transverse to the direction of propagation of the wave motion. These waves vary in length from approximately 3.85 to 7.60 ten-thousandths of a millimeter.

The impression of *color* produced when the light energy impinges on the retina of the eye varies in a complex way with the wavelength, the amplitude of vibration, and various other factors and conditions, some of which are beyond the scope of this text.

Waves of similar character, but whose lengths are above or below the limits mentioned in a preceding paragraph, are not perceptible to the average eye under normal conditions. The very short waves between 1.0 and 3.85 ten-thousandths of a millimeter (100 to 385 microns) in length constitute ***ultraviolet*** light and are detectable by photographic or chemical action. Those waves which are longer than 7.60 ten-thousandths of a millimeter (760 microns) are the ***infrared*** waves and are detected by their thermal (heat) effects.

The electromagnetic theory of light as set forth by Maxwell, an English physicist, holds that these waves, including those of light proper, are the same kind as those by which electromagnetic oscillations are propagated through ether, and that light is an electromagnetic phenomenon. The most important phenomena of light are reflection, refraction, dispersion, interference, and polarization.

Optics is the science pertaining to light, light's origin and propagation, the effects to which it is subject and which it produces, and other phenomena closely associated with it. *Geometrical optics* is concerned with the optical phenomena associated with reflection and ordinary refraction as they can be reduced mathematically from the simple laws of reflection which have been derived from observation and experimentation. *Physical optics* is concerned with the description and explanation of all optical phenomena in terms of physical

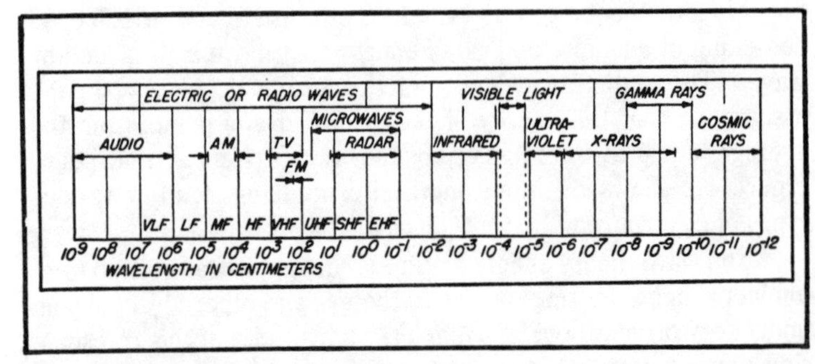

Fig. 4-2. Electromagnetic wavelengths.

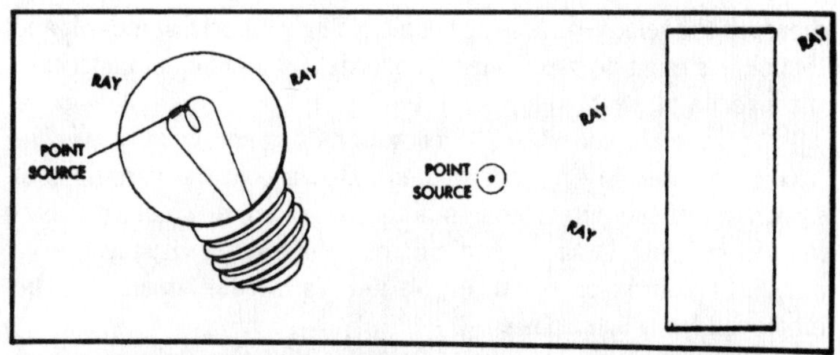

Fig. 4-3. Light rays from a point source.

theories, such as wave theory in general, electromagnetic phenomena, quantum mechanics, and other light properties.

One of the mysteries pertaining to radiant energy such as light, heat, and electromagnetic waves is the medium through which it is conducted. Only because the medium absorbs the energy and changes it to some other form are we able to recognize its existence and determine its characteristics. As indicated earlier, the word ether is used to name the medium through which radiant energy is conducted. But it is not known what the medium actually is.

The most obvious fundamental property of light is that it travels in straight lines when passing through a homogeneous (uniform in density and composition) medium. The rectilinear (straight-line) propagation of light supports the idea that a ray of light is the rectilinear path in a homogeneous medium along which light is propagated or transmitted.

By choosing one point on a luminous body and from that point drawing a straight line in the direction of the propagation of light, you can represent a ray of light. From this point source of light, you can draw an infinite number of rays, as shown in Fig. 4-3. This collection of rays, or cone of light, is referred to as a *pencil of rays*. Two rays from such a group are enough to locate the point source of the light by simple geometric means. The point source is the point of intersection of lines extended along the paths of the rays.

UNITS OF LIGHT INTENSITY

Luminous intensity or brightness of light represents the degree to which visible light is present in the radiant energy emitted by the source.

The retina in the human eye is sensitive to a relatively small portion of the radiant energy emitted by an incandescent body. To measure the relative intensities of visible light, special standards and techniques must be used. Such standards and techniques constitute the science of *photometry*.

Light can be considered as a flow of radiant energy or ***luminous flux*** (expressed in ergs per second). Because of the variations of sensitivity of the human eye to different colors, or different wavelengths of light), luminous flux cannot be measured visually in ergs per second. In place of the erg, a unit called the lumen must be employed for this purpose.

A ***lumen*** is the amount of light flux radiating from a uniform one-candela (1 cd), or one-candle, source throughout a solid angle of such size as to surround a unit area at a unit distance from the source.

Light flux, or ***luminous flux,*** refers to the rate at which a source emits light energy, evaluated in terms of its visual effect.

If you imagine the one-candela source to be located at the geometric center of a hollow sphere having a radius of one foot, then each square foot of the interior surface of the sphere receives one

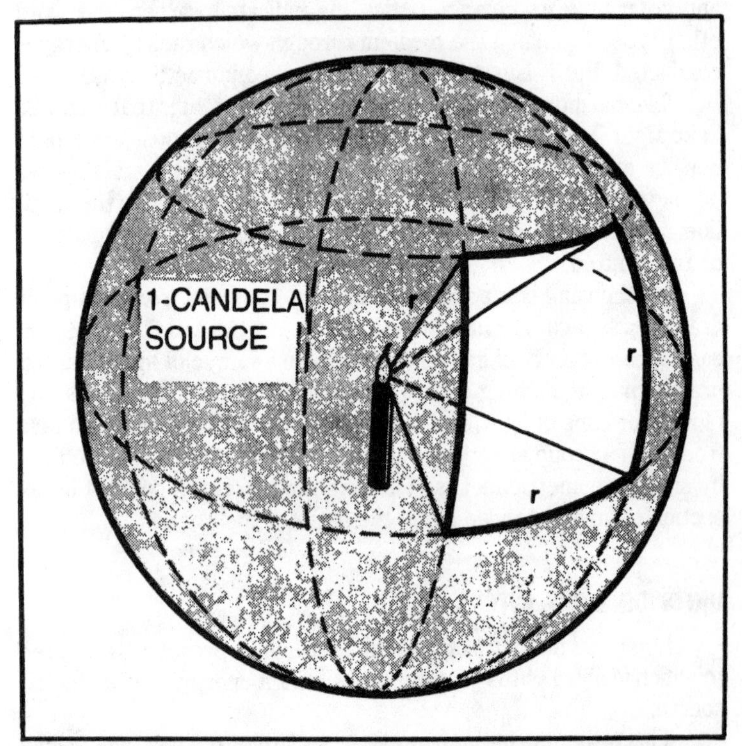

Fig. 4-4. Light falling on an area R^2 at distance r from 1 candela source equals 1 lumen.

lumen of light. Since the total area of the sphere is four pi (4π) square feet, the total light emitted by the one-candela source is 4π lumen. Figure 4-4 of a sphere and solid angle illustrates this example. Light from the one-candela source falling upon a unit area at unit distance represents one lumen.

Most sources of light have different luminous intensities along different directions. The average of the candelas measured in all directions about a source of light is called the *mean spherical candela*. If a source having a mean spherical candela of 1 cd emits 4π lumens of luminous flux, the total flux, F_T, in lumens emitted by a source of mean spherical candela, I_O, can be expressed by the equation $F_T = 4\pi I_O$.

The primary standard of luminous intensity, developed by the National Bureau of Standards, consists of a glowing enclosure operated at the temperature of solidifying platinum (2046° Kelvin or 2187° Celsius) and arranged as shown in Fig. 4-5. The platinum is contained in a crucible of fused thorium oxide or quartz, surrounded by a heat-insulating material. The unit is placed in an alternating magnetic field so that the platinum in the crucible is melted by the currents induced within it. A viewing tube of fused quartz or thorium oxide, containing some finely powdered thorium oxide, is enclosed by the molten platinum and serves as a "black body" radiator. The brightness within this tube is considered to be 60 candela per square centimeter (60 cd/cm^2) when the metal, in cooling slowly, attains its solidifying temperature. The *standard candle*, or *candela*, is one-sixtieth of the luminous intensity of one square centimeter of a hollow enclosure at the temperature of solidifying platinum.

The amount of luminous flux which falls upon a surface and the area of the surface together determine the degree of illumination. The illumination is uniform only when a single source of light is employed and when all points on the illuminated surface are equidistant from the source.

A light intensity of one lumen per square foot is a foot-candela (foot-candle) and is the amount of illumination received on a surface one foot distant from a standard candela (candle). The illumination of a surface is measured by the number of lumens incident upon a unit of area:

$$E_T = I_C / d^2$$

where E_T represents the total illumination, I_C equals the intensity of the source in candela, and d equals the distance from source to surface.

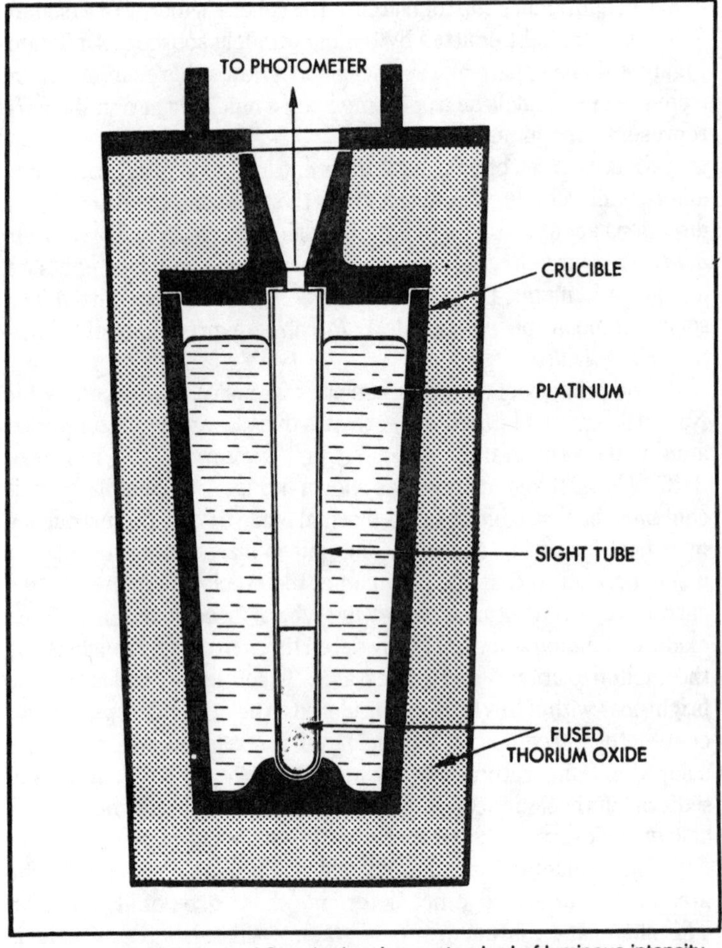

Fig. 4-5. National Bureau of Standards primary standard of luminous intensity.

If you consider a surface of area, A, as receiving a total luminous flux, F_T, you can express the illumination of the surface in terms of lumens per unit of area, such as lumens per square foot. Expressed as an equation,

$$E_T = F_T/A$$

The degree of illumination which a light source produces upon a given surface depends upon the intensity of the source and its distance from the surface, provided that the rays of light pass through a uniform medium and strike the surface normally.

Increasing the intensity of the source produces a proportional increase in the light flux falling upon the surface. Increasing the distance of the surface from the source decreases the illumination of the surface by an amount proportional to the square of the distance. Doubling the distance between the source of light and the illuminated surface will reduce the illumination of the surface to one-fourth its original value.

This effect is common to all forms of radiant energy and is expressed in the *inverse square law*, which states that *the radiant flux density at any surface varies inversely as the square of the distance of that surface from the source of radiation*.

The numerical value of illumination is identical whether expressed in foot-candelas (foot-candles) or in lumens per square foot.

Remember, the equation $E_T = I_C/d^2$ can be used to calculate illumination only for a spherical surface with the source of radiation at its center. This condition is known as *normal illumination.*

For small surfaces where the distance from source to surface is large in comparison to the dimensions of the surface, the formula can be used with little error. Over large surfaces where all the flux paths are not perpendicular to the surface, the diffusion of light is not uniform. The flux intensity is reduced at points farther removed from the source by an amount proportional to the cosine of the angle of incidence (i). This factor must be included in calculations. The formula

$$E_T = \frac{I_C \cos i}{d^2}$$

is much more accurate under these conditions. Even with this formula the surface dimensions must still be small compared with the distance from the source.

Up to this point, only the luminous intensity of a source in terms of point sources has been considered. When referring to luminous intensities of larger surfaces which may or may not be self-luminous, the quantity, or term brightness, must be used to specify luminous intensities of unit area. *Brightness* is defined as the luminous intensity of a unit area of a surface in a given direction. Brightness is expressed in terms of candlepower per square unit of area. It is generally expressed in terms of square centimeters.

The difference between brightness and illumination can be illustrated by considering this page which you are reading. The page is uniformly illuminated (or nearly so), but the printed letters reflect less of the incident light and therefore are less bright than the white paper upon which they are printed.

Brightness of a surface and the illumination on the surface would be numerically equal only if the surface reflected all of the light that fell upon it. The information to follow shows the approximate values of brightness of some familiar self-luminous and nonluminous objects.

Sun's disk.	150,000,000 cd/ft^2
Crater of a carbon arc.	14,000,000 cd/ft^2
Tungsten lamp filament.	450,000 cd/ft^2
Moon's disk.	450 cd/ft^2
Clear blue sky.	370 cd/ft^2
Newspaper.	1.8 cd/ft^2

Generally the brightness of a surface depends upon the direction from which it is viewed, but there are some materials which scatter light in such a manner that their brightness is the same from all angles of view. Examples of such light-diffusing substances are magnesium oxide and new-fallen snow. For surfaces of this type, a unit of brightness called the *lambert* is used.

A lambert represents the brightness of a perfectly diffusing surface which is emitting or scattering light in the amount of 1 lumen per square centimeter. When the reflected light is less than the incident light, the brightness in lamberts is equal to the product of the illumination and the reflection coefficient of the surface material.

MEASUREMENT OF LIGHT INTENSITY

The relative intensities of two or more sources are not discernible to the human eye by direct viewing, but whether or not two surfaces side by side are equally illuminated can be determined accurately. The matching of illumination on two adjacent surfaces is the basic principle of the *photometer*, a device employing two lamps located at some suitable distance apart with a screen located between them. Each side of the screen is illuminated normally by one of the sources. The flux paths from each source are perpendicular (or nearly so) to all points on the surface of the screen.

The screen is moved along the flux path between the two sources until the same degree of illumination is observed on both sides. The distances from the lamps to their respective sides of the screen are then measured. From the equation

$$E_T = \frac{I_C}{d^2}$$

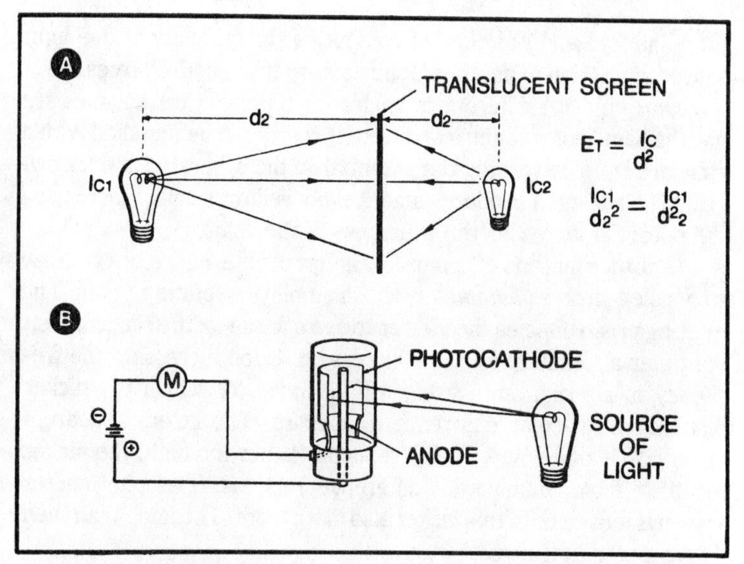

Fig. 4-6. Measuring light intensities. (A) Inverse square law method. (B) Photoemissive cell method.

you can establish the ratio

$$\frac{Ic_1}{d_1^2} = \frac{Ic_2}{d_2^2}$$

where Ic_1 and Ic_2 are the luminous intensities of the lamps in candelas and d_1 and d_2 are their respective distances from the screen. If the value of either Ic_1 or Ic_2 is known, the value of the other can be readily computed. Figure 4-6A illustrates the application of the inverse square law as applied in the formula.

If distance d_1 is found to be twice the value of d_2 and Ic_2 represents a source of 16-candela intensity, the intensity of Ic_1 must be four times that of Ic_2 to produce the same illumination of the translucent screen.

From the second equation above,

$$\frac{Ic_1}{d_1^2} = \frac{16}{d_2^2}$$

Therefore, if d_1 is equal to twice d_2 and if Ic_2 is 16 candela,

$$\frac{Ic_1}{2^2} = \frac{16}{1^2}$$

Thus, Ic_1 equals 4 times 16, or 64 candela.

The foregoing method of computing the intensity of one light source in relation to another is satisfactory when both sources are of the same color. A different procedure is necessary if the lamps are not the same color. Lamps of different colors can be matched with a standard lamp by varying the potential on the calibrated lamp. Lowering the potential produces a noticeably yellower color. Increasing the potential increases the whiteness of the color.

Another method of comparing lamps of different colors is to use a so-called *flicker photometer*, which employs a rotating prism. The rotating prism enables the viewer to see one side of the screen, then the other alternately in rapid succession. Upon increasing the frequency of alternation, a value will be found for which the flicker resulting from color difference disappears. The colors appear to blend into a single hue. If the frequency is not too high, the flicker resulting from illumination difference remains. The photometer screen is moved until this flicker also disappears. The lamps can then be compared in the usual way.

Photoelectric cells with suitable light filters can also be used to compare the candlepower of lamps or other sources of light of different colors.

Photoelectric current is proportional to the illumination on the photocathode, and the candlepower of the lamp under test can be expressed in terms of the current it produces compared with that produced by a calibrated standard lamp. Figure 4-6B illustrates how a photoemissive cell can be used in this application.

VELOCITY OF LIGHT

Early experiments to determine the velocity of light gave inaccurate results because of time losses in the operation of the equipment used. But recent observations using improved optical devices and electronic timing controls have greatly reduced the error. The velocity of light is now considered to be approximately 186,300 miles per second. In general calculations, either 300,000,000 meters or 186,000 miles per second is used as the velocity of light and radio waves.

Modern methods for calculating the velocity of light usually employ some modified version of the rotating-mirror method used by the French physicist, Foucault, in 1850. Foucault's method involved the directing of a narrow beam of light upon a plane mirror rotating at high speed. A fixed mirror located at a considerable distance from the flashing (rotating) mirror received the momentary flash of light and reflected it back to the rotating mirror from which it was again reflected. During the interval of time required for the light beam to travel through the measurable distance between the rotating mirror and the stationary mirror and back, the flashing mirror rotated through a definite angle.

From the angular velocity of the rotating mirror and the distance between the mirrors, the time required for the light to travel from the flashing mirror to the fixed reflector and return was computed. From these factors the velocity of the light was determined.

Foucault's method was improved by an American physicist, Michelson, who used a rotating octagonal mirror. Light from a source of high intensity was reflected from one face of the rotating mirror to a distant plane mirror. The returning light ray from the fixed mirror would strike another face of the rotating mirror from which it was reflected into the observer's telescope.

REFLECTION AND REFRACTION OF LIGHT

Two of the most important phenomena of geometric optics are reflection and refraction. Each of these phenomena is characterized by a basic relationship or natural law and occurs in all optical systems such as the human eye, lenses, prisms, telescopes, microscopes, etc.

Law of Regular Reflection

The fundamental law of regular reflection states that when a ray of light is reflected from a surface, the angle of reflection is equal to the angle of incidence. The reflected ray, the incident ray, and the

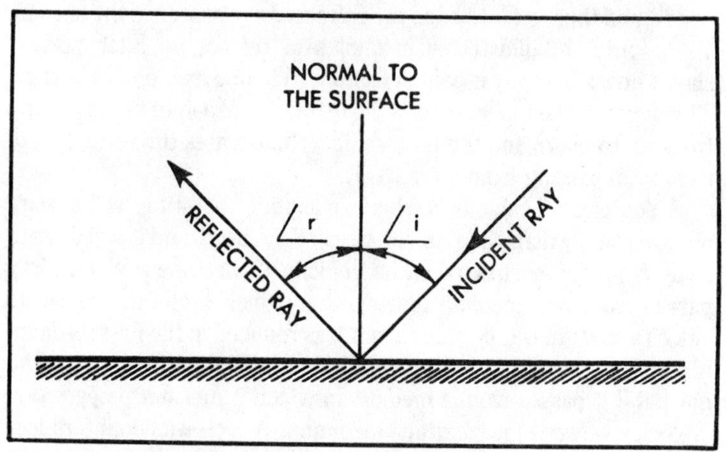

Fig. 4-7. Angle of reflection equals angle of incidence.

normal (a line perpendicular to the reflecting surface at the point of reflection) all lie in one plane. Figure 4-7 illustrates this law.

The law of reflection applies when light is reflected at the interface, or common boundary, between two unlike media such as air and a solid surface. The one medium, air in this case, is transparent and the other, a solid, is opaque.

Refraction of Rays

If the second medium is not opaque, some of the light passes on through it and is refracted or bent in a direction which generally is different from the direction of the incident rays. The refracted rays still lie in the same plane as the normal to the surface. The degree to which this bending action takes place determines the ***refractive index*** of the medium.

Each refracting medium has a specific refractive index in any one refracting medium. There is a constant ratio between the sine of the angle of incidence and the sine of the angle of refraction in any one refracting medium. Both angles are measured with respect to the normal, in this case a line, perpendicular to the reflecting surface at the point of incidence and refraction.

According to Snell's law, a ray is bent toward the normal if the second medium has a greater refractive index that the first, and is bent away from the normal if the second medium has a smaller refractive index.

The angle of refraction is smaller for some mediums than others. For example, the angle of refraction is smaller for glass than for water because the refractive index of glass is greater than that of water, and thus tends to bend the refracted ray nearer to the normal.

Figure 4-8 illustrates examples of refraction. Each portion shows how a light ray is bent in passing from one medium to another. The figure at the top illustrates the refraction of a light ray in passing from air to glass, and the bottom figure illustrates the refraction of the ray in passing from glass to air.

You can conclude from this explanation that a ray will deviate toward the normal when its velocity is decreased, and it will deviate away from the normal when its velocity is increased. When a ray passes from one medium into another which tends to reduce its velocity to a greater degree than it was reduced in the first medium, it deviates toward the normal. The ray deviates away from the normal if it passes from a medium in which it met more opposition (velocity slowed) into another medium in which it encounters less opposition.

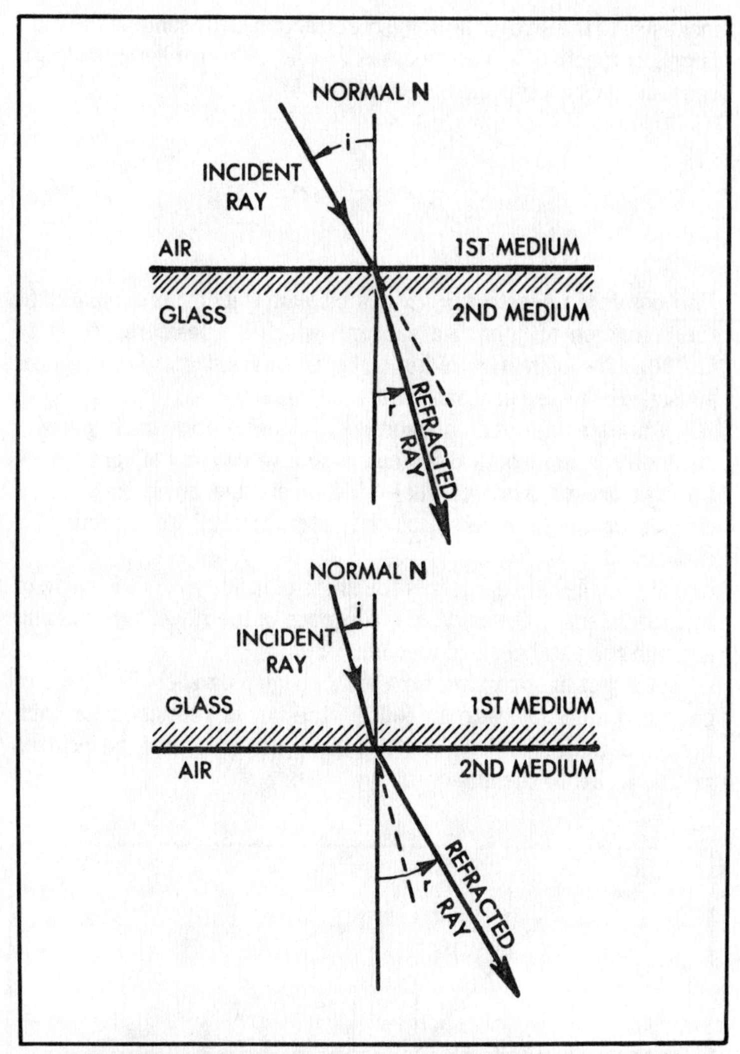

Fig. 4-8. Refraction of light rays.

As noted above, the ratio of light velocities in two mediums which are in contact is a constant for those two mediums. As noted above, this ratio is referred to as the *refractive index* of the second medium with respect to the first. The refractive index for two mediums is represented by the symbol $\mu_{1,2}$ (mu sub 1, sub 2), with the order of the subscripts indicating the direction of the ray of light.

The *law of refraction* states that when a wave travels obliquely from one medium into another, the ratio of the sine of the angle of incidence to the sine of the angle of refraction is the same as the ratio of the respective wave velocities (v) in these mediums and is a constant for two specific mediums. Thus,

$$\frac{\sin i}{\sin r} = \frac{V_1}{V_2} = \mu 1{,}2$$

The ***absolute refractive index*** of a medium is its index compared to the refraction of light in a vacuum, which is considered as unity (1.000). The refractive index of air is so small that for practical purposes it is used as a standard.

Another important phenomenon of refraction which you will encounter in some optical systems is that when a ray of light passes through one or more parallel-sided media and emerges into the original medium, it is displaced laterally but its direction is unchanged.

In Fig. 4-9, i represents the angle of incidence, r the angle of refraction, and e the angle of emergence of the ray of light passing through the parallel-sided medium, glass.

A ray of light passing from air through a parallel-sided pane of glass and emerging into air follows the law of refraction for each surface of the glass. Thus, angles i and e are equal, and the incident ray is parallel to the emergent ray. Also,

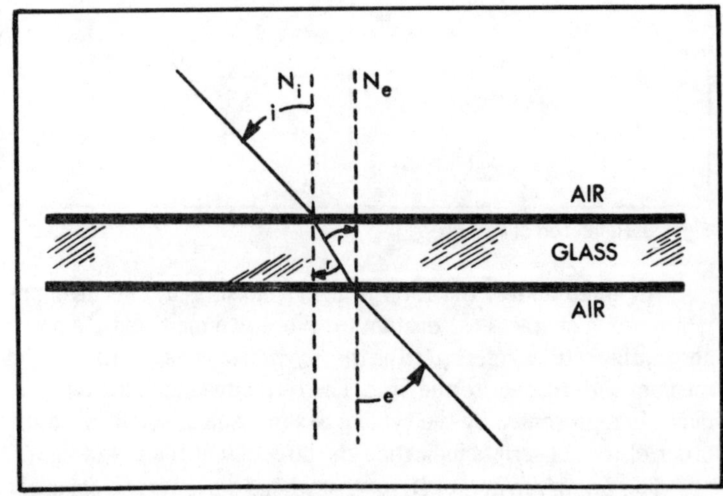

Fig. 4-9. Passage of light from air to glass to air.

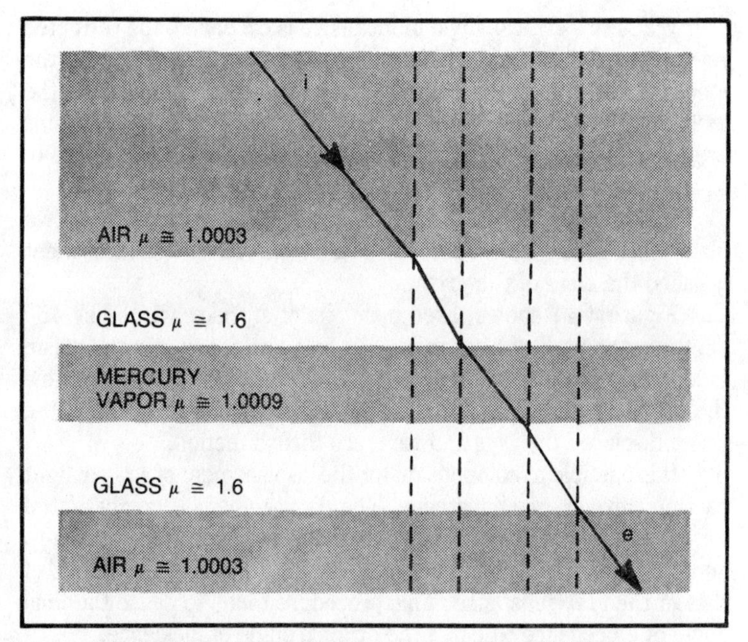

Fig. 4-10. Refraction of light passing through several parallel-sided media.

$$\mu_{a,g} = \frac{1}{\mu_{g,a}}$$

which shows that the refractive index of glass, *g*, with respect to air, *a*, is the reciprocal of the refractive index of air with respect to glass.

These same principles hold true for a ray of light passing through more than one parallel-sided medium, as illustrated in Fig. 4-10. The figure shows how light is refracted in passing through several parallel-sided media. The refractive index of each medium is noted in the figure. Notice in the illustration that the angle of refraction in the glass is less than in the air or gas.

A ray of light, passing through a medium with a high refractive index toward one with a lower refractive index, is refracted if the angle of incidence is not too large when it passes into the second medium. If the ray is inclined to an ever-increasing angle of incidence, it arrives at some position at which it no longer passes into the second medium. Instead, it is totally reflected at the common surface of the two mediums. Such a condition is known as the *critical angle of incidence*. The critical angle of incidence is the maximum angle at which the ray of light striking the surface of a medium passes through it.

When this critical angle of incidence is exceeded, the refracted ray grazes the surface with an angle of refraction of 90° in the medium, and the ray is totally reflected from the surface of the medium. This principle is often applied in optical instruments. Prisms are used to achieve total reflection whenever avoiding the use of silvered mirrors is desired.

Basically, a prism is a transparent body bound in part by two plane faces which are not parallel. The line in which these faces meet is called the edge of the prism.

Figure 4-11 shows three positions of a prism, which has 45° angles in the path of light rays. The first position shows the prism functioning as a plane mirror tilted, the second position shows how the prism inverts an image, and the third position shows how the prism displaces the rays and reverses their direction.

It is possible to compensate for the displacement of a ray of light passing through one substance. The displacement is compensated for by causing the ray to emerge from the displacing substance into another material which has a refractive index that is the reciprocal of that of the first substance. This procedure tends to cause the final angle of emergence to equal the original angle of incidence.

Variations in atmospheric conditions produce variations in the refraction of light from celestial bodies; therefore, in selecting a star as a fix for a celestial navigation system, we choose one whose light has an angle of incidence as nearly equal as possible to the normal for the earth's atmospheric layer.

Light rays passing through successive layers of air of varying densities or through translucent media with varying indexes of refraction are bent or directed along a new path of radiation by each layer. When this process occurs at a visible rate, it produces an effect called *scintillation*, more commonly called glitter or twinkle. The twinkling of starlight is produced in this manner. A similar effect can be detected with respect to electromagnetic waves passing through

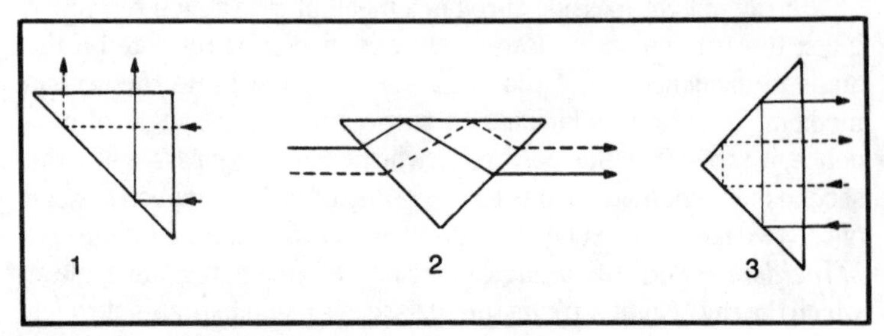

Fig. 4-11. Total reflecting prism in various positions.

zones of conductance which vary in refractive characteristics for each specific wavelength of the radiations.

Light reflected from surfaces with many facets, such as cut diamonds, tends to produce the effect of flashing or scintillating. Electromagnetic waves, reflected from various objects and arriving at the receiver at intervals slightly later than the direct beam, tend to produce phase differences and frequency in varying degrees. This results in phase cancellation and interference, referred to as fading, swinging signal, or ghosts.

In television, a ghost image on the picture screen occurs when the transmitted signal arrives at the receiver along more than one path, the paths being unequal in distance. This condition produces two or more images slightly displaced in phase and time.

Radar signals are subject to the same effect. The effect is present when the signals are reflected from moving surfaces or from more than one object, or when they are refracted by inequalities in terrain or atmosphere. To minimize this effect of fluctuation or scintillation, the radar beam is narrowed so that both the incident and reflected beams travel through a narrow zone and are less subjected to reflection from objects other than those directly along the principal axis of the beam. When scintillation is produced by a fixed or constant succession of media, the frequency of the scintillation can be used to identify the path or locate the source of the radiation.

Spectra

The rainbow of color produced by sunlight passing through a crack and diffracting medium into a darkened room is a visible color spectrum. This phenomenon is the result of refraction and dispersion of light rays, and it exemplifies the basic principle of spectrography.

White light is composed of light rays of many hues blended together. When such light is passed through a narrow opening or slit into some diffracting medium, rays of each different color are diffracted at different angles and spread out into merging bands of six principal colors. These colors are red, orange, yellow, green, blue, and violet. The diffracting of radiations of light at different angles produces a rainbow or spectrum of light. The prism shown in Fig. 4-12 is acting as a diffracting medium, thus producing a spectrum of light.

A spectrum is composed of hundreds of hues which are grouped broadly into the six principal colors mentioned above. These colors appear in the order of increasing deviation from red to violet.

The color of light is determined by its frequency of vibration,

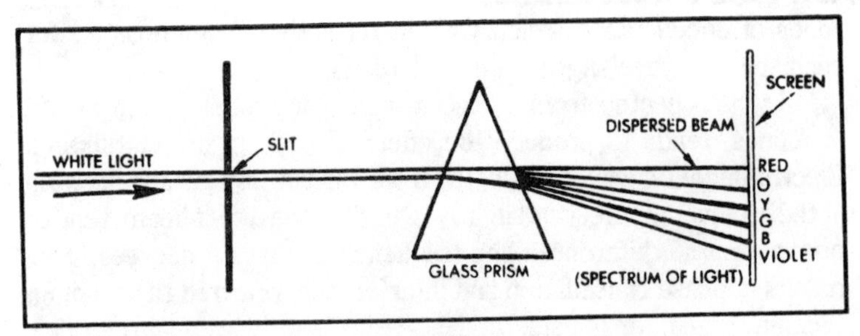

Fig. 4-12. Dispersion of light through a prism.

the frequency being lowest for red and greatest for violet. The wavelength, λ (lambda), is longest for infrared light and shortest for ultraviolet light, and each color possesses certain characteristics and properties which can be used in light-sensitive or other optical systems.

The deviation produced by a prism is greater for the higher frequency components of white light than for the lower frequency hues, but no definite relationship exists between frequency and deviation. Prisms of different materials spread the component colors of the spectrum in varying degrees.

Some materials produce anomalous dispersion; that is, some prisms are composed of materials that do not disperse white light in the regular sequence of colors. Such prisms deviate certain colors to a much greater degree than others and also absorb certain portions of the spectrum.

In a quantitative study of the spectrum, it becomes necessary to refer to each particular part of it with definiteness. This is done by specifying any hue by the vibration rate of the light source and its corresponding wavelength. As in all other forms of wave motion, the velocity of light is equal to the wavelength times the frequency,

$$v = f\lambda$$

where f equals the frequency and λ the wavelength. When light is retarded by a medium such as glass, the frequency is unaltered. Since velocity decreases, λ must also decrease in direction proportion. Therefore, λ is not a constant quantity for a given vibration, but it depends upon the medium.

In general, values of λ for different colors of a spectrum are given as the wavelength in air or in a vacuum. Wavelengths are so short for visible light that a special unit of length shorter than the centimeter is commonly employed to measure them. This unit, as stated previously, is known as the ***angstrom***, named in honor of a Swedish physicist. The angstrom unit, Å, is equal to one-hundred millionth of a centimeter (10^{-8}cm).

In some cases of lower frequency radiations, the micron, μ, can be used as a measuring unit. One micron is equivalent to 10,000Å.

Spectra are often classified into three general types: emission, absorption, and solar spectra.

Emission Spectrum. A spectrum produced by a glowing object is termed an emission spectrum. Its appearance depends primarily upon the composition and state of the luminous object.

Incandescent solids and liquids produce continuous spectra, extending from color to color without interruption. Luminous gases and vapors yield spectra consisting of definitely placed bright lines. Each bright line is an image of the slit through which the radiation is received.

Every gas emits radiation of particular wavelengths, and each spectrum is characteristic of the radiating substance. For example, sodium vapor yields two bright lines in the yellow part of the spectrum, while mercury vapor yields several bright lines, the most conspicuous being in the green and blue regions. Figure 4-13 shows the bright-line spectra of several elements over the range of the visible spectrum. The continuous spectrum at the top is an uninterrupted series of images of the illuminated slit.

The number of lines in a bright-line spectrum depends upon the amount of energy with which the atoms of the source are excited to glowing, as well as on the nature of the source. The greater the excitation of the atoms of a substance, the greater the number of lines that appear in its spectrum.

Absorption Spectrum. Absorption spectra occur when white light has passed through an absorbing medium before the light is dispersed. A glowing solid, or other source from which the light radiates, yields a continuous spectrum, but when the light passes through the absorbing medium, radiations of particular wavelengths are absorbed. The resulting spectrum is usually crossed by dark spaces because of the absence of the absorbed radiations. If the absorbing material is solid or liquid, these dark spaces appear as broad, structureless bands. If the material is gaseous, the dark spaces consist of dark lines which occupy the same positions as the bright lines in the corresponding bright-line spectrum.

Solar Spectrum. A spectrum formed by radiations from the sun is called the solar spectrum. This spectrum appears continuous from a casual inspection. A more critical examination shows that it is

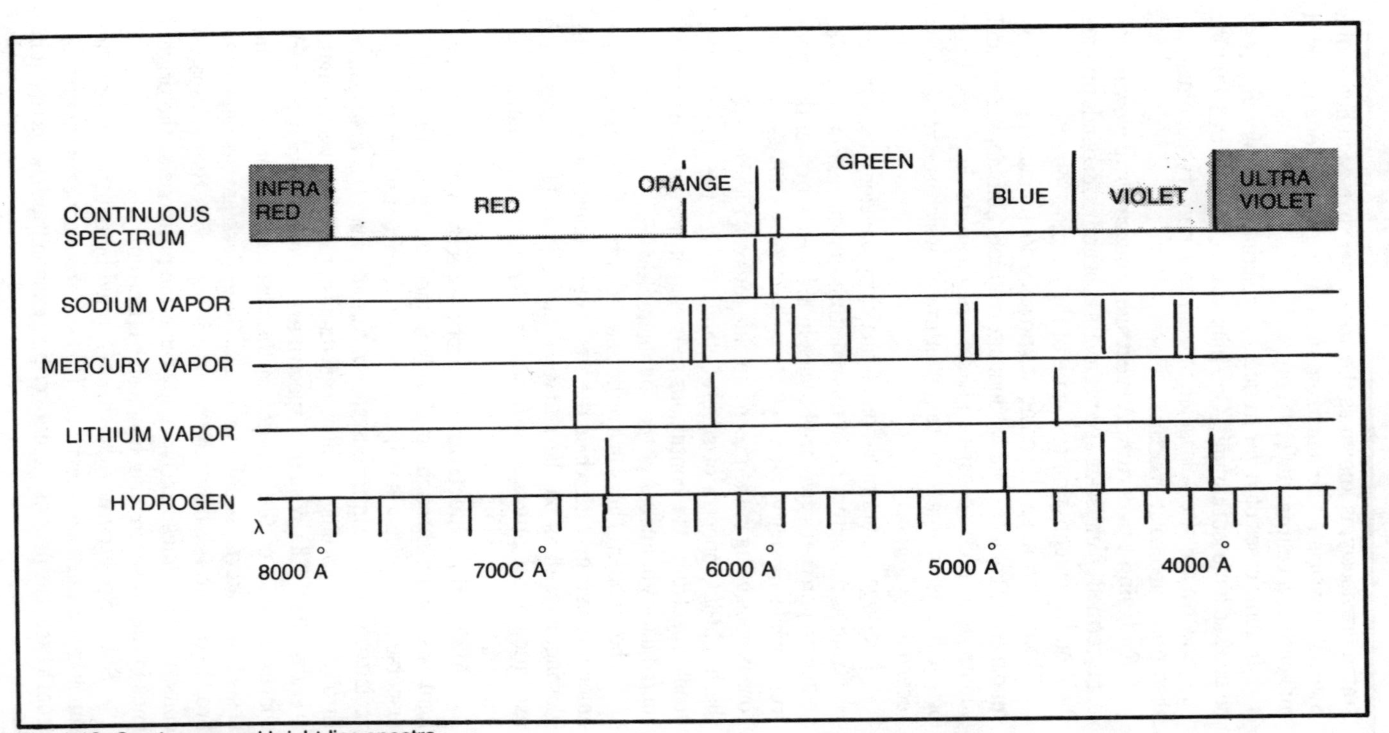

Fig. 4-13. Continuous and bright-line spectra.

crossed by numerous dark lines. No doubt the sun's radiation comprises all wavelengths in the visible range, but in passing through the sun's atmosphere, certain wavelengths are absorbed. The spectrum observed is in reality an absorption spectrum of the sun's atmosphere.

Continuous spectra (Fig. 4-13) are produced by light from incandescent solids and liquids. Bright-line spectra come from incandescent vapors or gases. Absorption spectra are produced by light passing from an incandescent solid or liquid through an incandescent vapor or gas.

POLARIZATION OF LIGHT

When the question arises as to whether light waves are longitudinal like sound waves or transverse like elastic waves, it is advisable to consider the phenomenon of polarization of light.

If a beam of light is passed through a substance composed of two colors, such as a crystal of tourmaline or a sheet of Polaroid, the beam's passage is restricted to a particular plane of vibration. The beam is said to be ***plane polarized*** and will pass through a second crystal of tourmaline or sheet of Polaroid only if the tourmaline or sheet is oriented exactly the same as the first. If the second crystal or sheet is rotated 90° with respect to the first, no light passes through.

The first crystal or sheet is called the ***polarizer***, and the second is the ***analyzer***. The analyzer is so named because the angle of polarization can be determined from the angle through which the analyzer must be rotated between the points of light passage and cutoff. Figure 4-14 illustrates the principle of polarization of a transverse wave. The vibrating rope corresponds to a beam of light in this analogy.

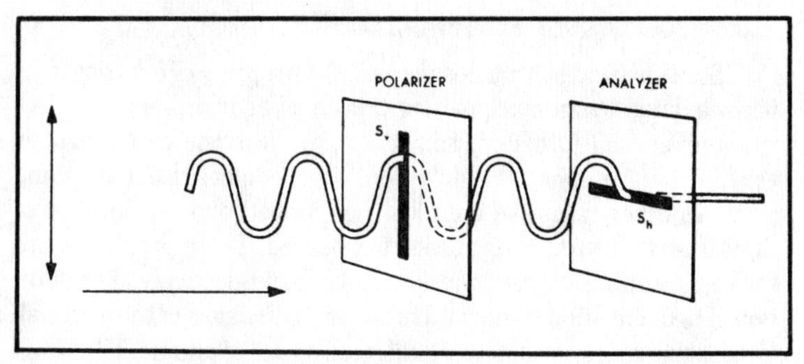

Fig. 4-14. Polarization of a transverse wave.

Vertical vibrations of the rope pass through a vertical slot (S_v), but they are stopped by the horizontal slot (S_h). The wave represented by the rope is polarized. It is vibrating in one plane only (in this example the vibrations are in the vertical plane). Light can be similarly polarized by certain optical substances such as tourmaline crystals and Polaroid, which is the tradename for a commercial product possessing this property.

If slot S_h were rotated 90° in the foregoing analogy, the waves would pass through it. Slot S_v limits the wave motion to the vertical plane and is the polarizer; S_h is the analyzer. Since S_h must be rotated 90° to permit the waves to pass, it is evident that the degree of polarization produced by S_h is the analyzer. Since S_h must be rotated 90° to permit the waves to pass, it is evident that the degree of polarization produced by S_v is 90°. The degree could be determined by the required distance of rotation of S_h between points of passage and cutoff, even if S_v were not visible.

Both theory and experiment have shown that longitudinal waves cannot be polarized. We can conclude that light must have a transverse wave motion if it has the form of wave motion at all.

Light can be polarized to a considerable extent by reflection alone. All reflected light is polarized to a certain degree because of scatterings caused by dust and vapor particles. Scattering is more apparent in light of short wavelengths. This statement is supported by the color changes of the sky. The midday sky appears blue, but a sunset appears reddish. The sunset appears reddish because the light travels through a longer path of earth's atmosphere, resulting in the short wavelength blue light being dissipated by scattering. This scattering leaves a predominance of the longer wavelength reddish hues.

FLUORESCENCE AND PHOSPHORESCENCE

Some natural substances possess the property of emitting light when excited by an external force, such as bombardment by electrons or certain forms of radiant energy. In some cases, light is emitted by the substance only while the bombardment is taking place. In other cases the emission can persist for some interval of time after the external excitation has ceased. These properties are termed *fluorescence and phosphorescence*, respectively. The duration of light emission is referred to as the *persistence* of the material. Phosphorescence is commonly referred to as *afterglow*.

One of the fluorescent substances used in electronic application is willemite (zinc orthosilicate or Zn_2SiO_1), a crystalline zinc compound varying in color from white, greenish-yellow, and green, to shades of red and brown. The white and green varieties are frequently used in the coatings on the screens of cathode-ray tubes used in oscilloscopes and radar indicators. Willemite is used in television kinescopes in combination with other substances which impart the desired degree of persistence.

Generally when the emission is induced by some form of radiation, the luminescent substance emits light which is of a longer wavelength than the incident radiation. This phenomenon is encountered in the fluorescent lamp, in which certain phosphorus compounds are excited by ultraviolet radiation, which is invisible. The compounds then emit visible light of various colors. The fluorescent compounds used in the "soft white" and "daylight" lamps are combinations of zinc beryllium silicate and magnesium tungstate.

Infrared radiations reflected from objects onto a photosensitive surface which is sensitive to infrared light can be used for seeing in the dark. A device called the snooperscope was developed during World War II for this purpose.

Essentially, the snooperscope consists of a source of infrared light and an image tube having a light-sensitive cathode. An infrared lamp projects a beam from which all visible light is filtered out. The reflected rays from the beam are caught upon a cesium cell, which is highly sensitive to infrared light and which forms the cathode of the image tube. The reflected radiations incident upon the cell cause it to emit photoelectrons which are then focused on a fluorescent screen to form a visible image, as in the cathode-ray oscilloscope.

You will find that a knowledge of the emission spectra of radiating surfaces is of value in understanding laser spectroscopy. This knowledge also is of value in understanding navigation systems employing light from certain fixed stars as a reference.

Such systems must be highly sensitive and selective and thus require the use of optical components of the highest precision. Marked progress has been made in improving basic optical instruments and in adapting them to new applications and developing electronic systems such as lasers which employ the basic principles of optics. This progress has created systems which are extremely sensitive and selective with respect to heat, light, and other forms of radiant energy.

Many of the present-day aeronautical and space systems are of these improved types (see Chapter 17), and further research and development undoubtedly will produce revolutionary improvements in their accuracy and reliability as well as in their adaptation to industrial and commercial usage.

SIMPLE OPTICAL DEVICES

To use the various phenomena of radiant energy, simple optical devices such as mirrors, prisms, and lenses are employed either singly or in combinations to suit the complexity of the requirements.

Functions of Mirrors

A mirror is a polished or smooth surface that forms images by means of reflected light. When an object is placed before a plane mirror, a right-side-up image is formed. The image appears to be just as far behind the surface of the mirror as the object is in front of it. You encounter this simple phenomenon whenever you approach a mirror.

Your image appears to move toward you from a point to the rear of and equally distant from the mirror surface. This phenomenon conforms to the laws of reflection and is an important one in regard to optics.

Figure 4-15 shows a plane mirror with an image that is direct and vertical and appears to be as far behind the mirror as the object is in front of it.

Curved reflecting surfaces are designed to give variations in the apparent position and dimensions of the image by deviating a beam of light, causing it to be more or less converging when reflected than it was when incident upon the mirror.

Spherical mirrors are classified as concave or convex. A concave mirror (Fig. 4-16) has its reflecting surface on the inside of the

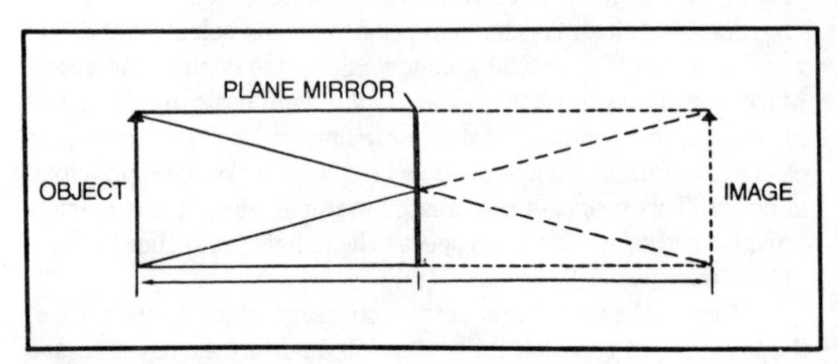

Fig. 4-15. Object distance equals image distance.

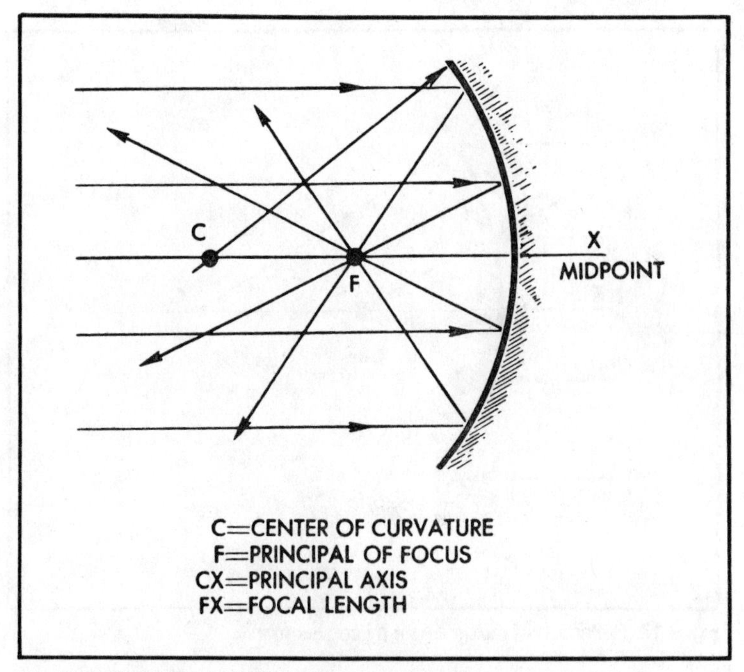

Fig. 4-16. Reflection of incident light by concave mirror.

spherical shell, while the convex mirror (Fig. 4-17) has its reflecting surface on the outer side of the shell.

A concave mirror is a converging reflector because it actually converges the light rays. A convex mirror is a diverging reflector because it makes the light rays appear to diverge.

The center of the spherical surface is called the *center of curvature* of the mirror. A line from the middle point of the mirror surface to the center of curvature is called the *principal axis* of the mirror.

Point *F* is the principal focus of the mirror. The distance of the principal focus from the mirror is the *focal length*. The principal focus of a spherical mirror is located on the principal axis halfway between the center of curvature and the mirror surface.

The ratio of the dimensions of a mirror with respect to the radius of curvature is referred to as the *aperture* of the mirror. Most optical mirrors are of small aperture and produce only a slight inclination of the incident rays with respect to the principal axis.

The ratio of the size of the image to the size of the object is referred to as the *magnification* of a mirror. A spherical convex mirror always causes the image to appear reduced in size but right side up. A spherical concave mirror can be made to produce inverted

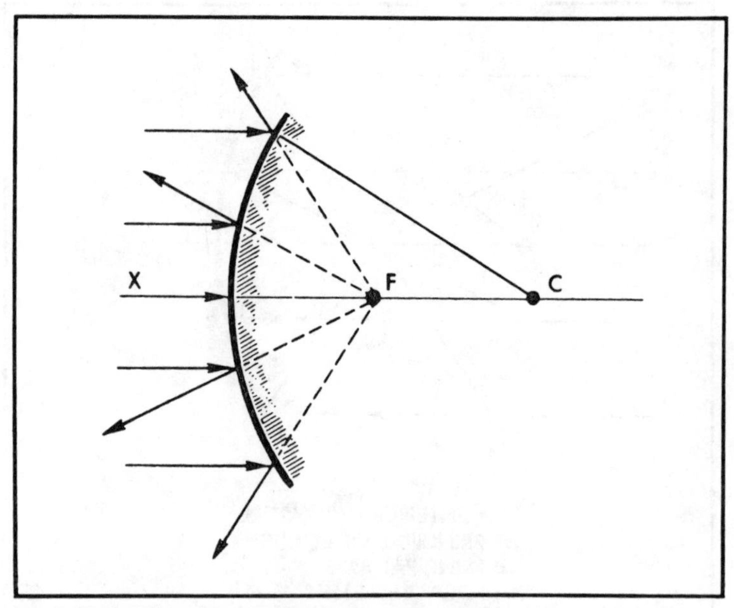

Fig. 4-17. Reflection of incident light by convex mirror.

images which appear to stand out in space or to produce right-side-up images which appear to be behind the mirror.

The inverted images are called *real images* because they appear to exist where the rays of light are focused and because they can be localized upon a screen. Right-side-up images are called *virtual images* and cannot be projected upon a screen.

Mirrors are part of the optical systems used in navigation and in surveying instruments. Often an optical system is used in conjunction with an electronic system which converts the optical data into electrical data. The electrical data can then be transmitted over great distances. Television is an example of such a combined optical and electronic system.

Uses of Prisms

Prisms are used in binoculars, spectroscopes, refractometers, and many other optical devices which use the phenomena of light dispersion and refraction.

The characteristics of light and the chemical composition of light sources can be determined by methods which involve the use of prisms.

In celestial navigation systems, the light from the reference star is passed through a prism in such a manner that either the predominant color or any desired color present in the dispersed beam can be used to activate a suitable photocathode or light-sensitive cell. The photocathode or cell produces and maintains a voltage or current output level which is proportional to the frequency (color) and intensity of the light which falls upon it.

Now you can visualize how such a constant voltage or current can be used to control a navigation system along a fixed path which is referenced to one or more specific light sources, such as stars. Many of the fixed stars emit light which is characterized by some specific color, such as Arcturus, which is orange, and Aldebaran, which is red. Spica is a spectroscopic binary; that is, Spica's white light is a blend of two predominant colors.

A star can be identified by its light spectrum once the spectrum has been tabulated. Spectroscopic equipment involves the use of prisms for the dispersion of the reflected light from the planet or the emitted light from the star.

Composition and Uses of Lenses

Usually, in addition to mirrors or prisms, lenses in some form are found in both optical and electronic-optical systems.

An optical lens is basically a piece of glass or other transparent material which has two opposite regular surfaces. Both surfaces can be curved, or one surface can be curved and the other plane.

Lenses are used singly or in combination with other lenses, prisms, or mirrors to perform specific functions. The primary function of a lens is to form an image by changing the direction of the rays of light. Such optical instruments as eyeglasses, cameras, microscopes, and telescopes are basically systems of lenses.

You will find that the curved surfaces of lenses generally are spherical, although in some rare instances cylindrical lenses may be encountered.

Spherical lenses can be classified broadly into six types as shown in Fig. 4-18.

1. Plano-concave: one plane surface and one opposite concavely curved surface.
2. Double-concave (biconcave): two opposite concavely curved surfaces.
3. Plano-convex: one plane surface and one opposite convexly curved surface.

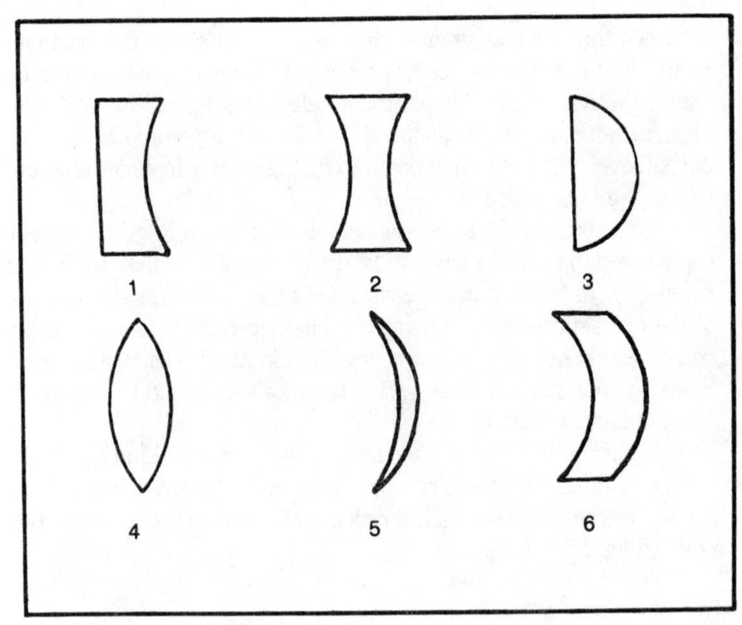

Fig. 4-18. Six types of spherical lenses.

4. Double-convex (biconvex): two opposite convexly curved surfaces.
5. Converging concave-convex (converging meniscus).
6. Diverging concave-convex (diverging meniscus).

In all of these spherical lenses, the line joining the centers of curvature of the two surfaces is a line of symmetry of the lens and is called the *axis of the lens*.

A lens whose focus for parallel rays is real is called a converging lens, as in the case of mirrors. And a lens which has a virtual focus for such rays is called a diverging lens.

Many of the principles of lenses which apply to light also hold true for other forms of radiation such as electromagnetic waves, cathode rays, etc.

The first outstanding scientific application of the principles of optics was in the field of astronomy. A telescope developed by the Italian physicist, Galileo Galilei, at the beginning of the seventeenth century is generally regarded as the forerunner of optical instruments; however, it is probable that eyeglasses were used prior to Galileo's time and most certainly some of the properties of simple lenses and prisms had been known for centuries.

The accuracy of optical/electronic systems depends on the quality of the lenses and other optical components and the precision of the mechanical construction and calibration.

Aberration of Rays Passing Through Lenses. The glass from which lenses are made must be free from impurities and must be homogeneous in chemical structure. The radii of the spherical surfaces must be properly selected. These factors are important in minimizing spherical aberration (deviation of rays from a focal point) to the greatest possible extent. Spherical aberration occurs when rays of light parallel to the principal axis of a lens do not all converge at a common focal point but intersect, instead, along the principal axis at various points. The intersecting of rays produces a blurred image.

A plano-convex lens, used so that the incident light falls upon its curved surface, produces little spherical aberration, as illustrated in Fig. 4-19.

Chromatic aberration is another problem encountered in the use of lenses. Chromatic aberration occurs when the various colors present in the incident light converge as individual points along the principal axis of the lens. This results from the variations in the refractive indexes of the colors. As shown previously in reference to the light spectrum, violet color has a greater refractive index than red; therefore, the focal length of the lens is less for violet light than for red. The violet rays converge at a point closer to the surface of the lens than the point at which the red rays converge. This phenomenon is illustrated in Fig. 4-20.

In passing through a lens, white light tends to disperse into its component colors which, in turn, tend to focus at different points

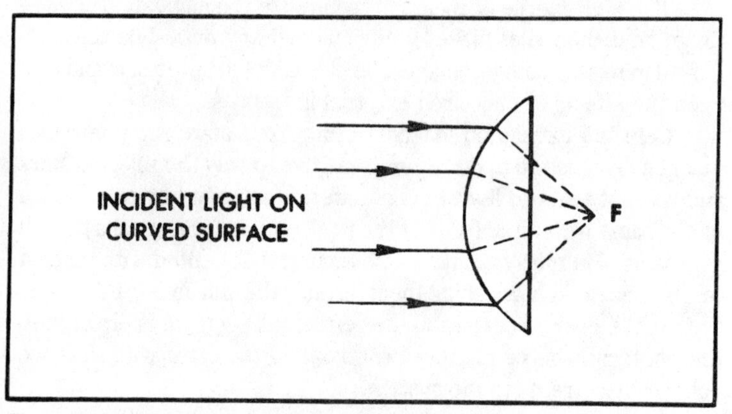

Fig. 4-19. Plano-convex lens with little spherical aberration.

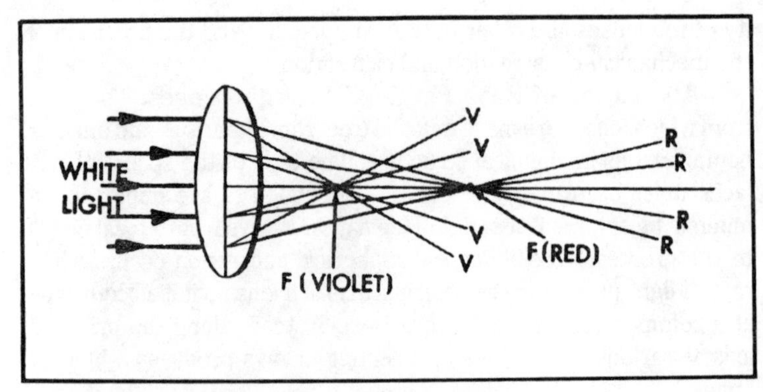

Fig. 4-20. Biconvex lens showing chromatic aberration.

along the principal axis of the lens. This phenomenon produces chromatic (color) aberration.

To overcome this effect, two or more lenses can be combined so that the divergence produced by one lens is nullified by the convergence of the other. Such a compounded lens is called an *achromatic lens* or doublet and generally consists of two lenses with opposite dispersion characteristics placed in contact with each other. Figure 4-21 illustrates a biconvex lens (positive), a biconcave divergent lens (negative), and two lenses combined to form an achromatic lens which is relatively free from chromatic aberration.

Indistinctness of an image produced by a spherical lens can result from *astigmatism* which becomes evident when rays of light pass through the lens obliquely and do not converge upon a common image point. This defect can be overcome by using two lenses suitably separated.

The high degree of accuracy required in the navigation of long-range guided missiles makes it obvious that any optical devices employed in missile navigation must be of the highest precision and free from the effects of undesired external influences.

Celestial navigation systems using fixed stars for references must be designed so that they are sensitive to only the selected stars and will not react to light rays of different intensity or color which might come within the fields of their telescopes. This means that all prisms used in the system must be accurately oriented with respect to the incident light reaching them through the starfinding telescope, so that the desired portion of the refracted spectrum is directed to the photocathode or photocell to maintain the standard reference voltage or current for the system.

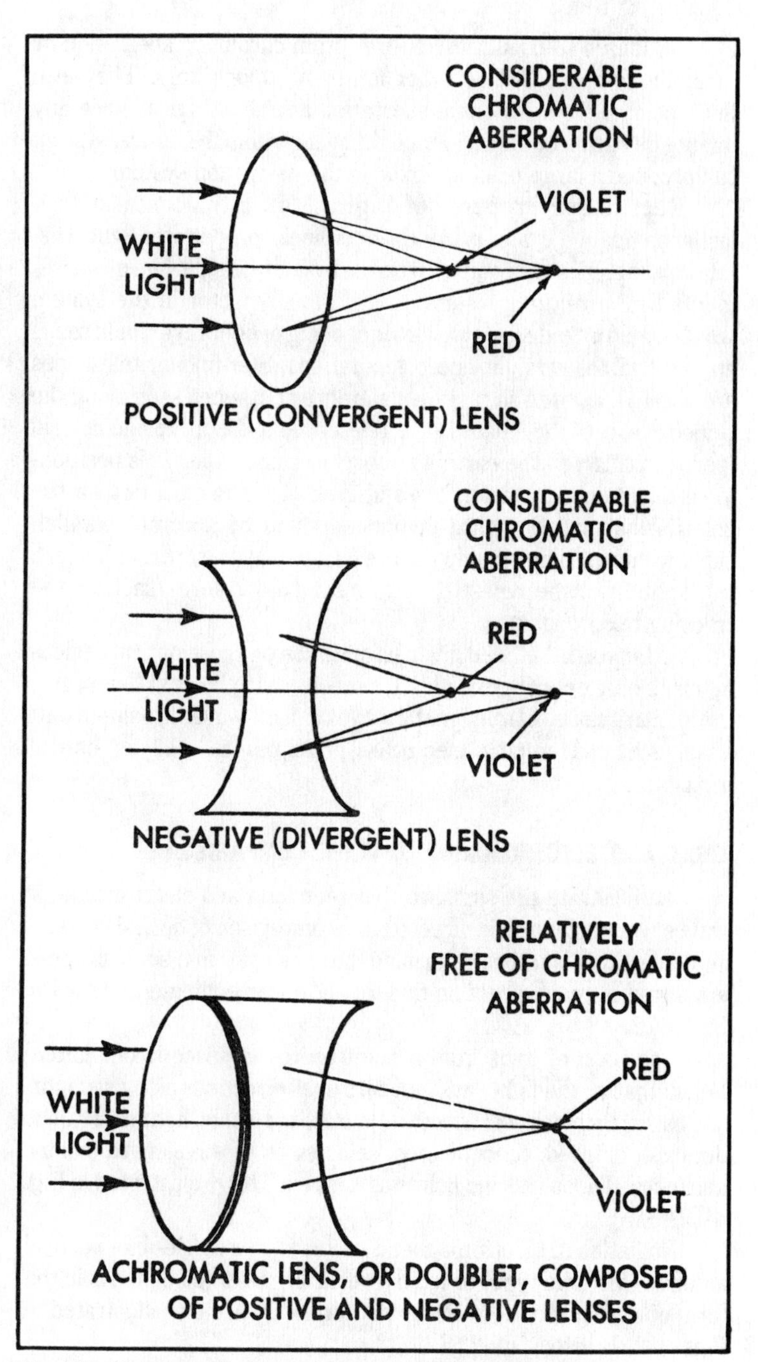

Fig. 4-21. Chromatic aberration caused by different types of lenses.

All lenses so used must be free from chromatic aberration or carefully compensated by other lenses or color filters. They also must be free from spherical aberration and astigmatism since any minute deviation of the incident light beam within the optical system can produce a large position error in the navigation system.

Rectilinear Property of Light. Most astronomical and celestial navigation systems use the rectilinear property of light. Distances and angles are computed on the basis of straight-line measurements to the reference stars; therefore, any factor in the system which would introduce deviation or bending of light rays would cause an error in the straight-line computations. Star-finding telescopes are usually mounted so that the reference star, when seen along the principal axis of the lenses in the telescope, is also in a plane normal (perpendicular) to the earth's atmospheric envelope. This positioning is maintained by means of a stabilized platform mounting for the celestial navigation optical components and by accurate parallel-sided windows. The windows serve as a passageway through which the light from the celestial body must pass before reaching the tracking telescopes.

So far you have been given information on those optical devices in which only optical principles are involved. In the next few pages, the similarity of electromagnetic waves to light waves is pointed out. Laser systems employ electronic principles as well as optical principles.

LIGHT AND ELECTROMAGNETIC WAVE COMPARISON

To illustrate the similarity between light and electromagnetic waves, your attention is directed to a comparison of optical systems and directional and beam-forming antenna systems, such as those employed in radio direction finders and radar equipment shown in Fig. 4-22.

Radiation of light from a point source in space is omnidirectional; that is, the light rays emanate in all directions along straight-line paths. By means of reflectors, lenses, and filters, light rays can be directed, diffused, concentrated, selected as to wavelength (color), and focused upon specific points as desired. This is illustrated in Fig. 4-22A.

Radiation of electromagnetic waves from a single-wire vertical antenna (Marconi type) or from a vertically mounted dipole is in the form of concentric fields along the horizontal plane as illustrated in Figs. 4-22B through 4-22E.

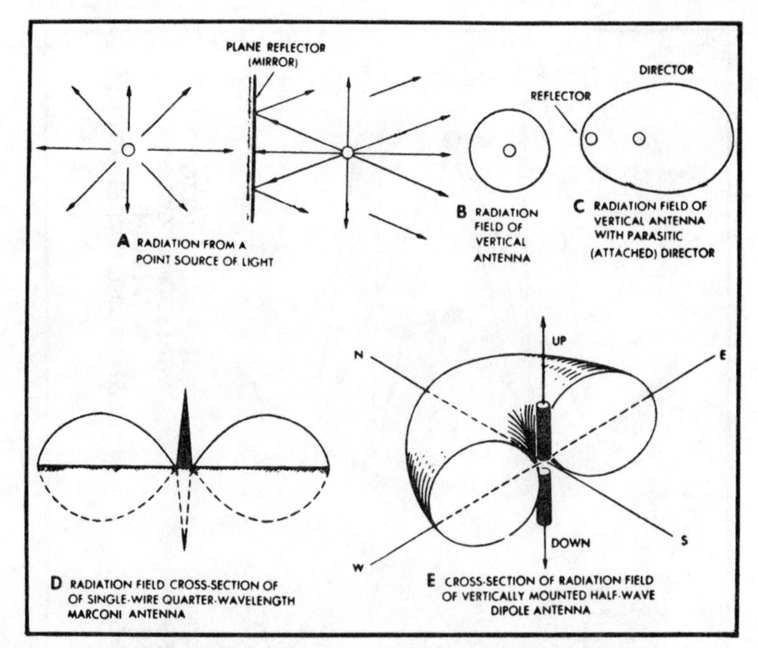

Fig. 4-22. Radiation of light waves and electromagnetic waves.

By means of suitable reflector and director elements, the electromagnetic radiations are directed and concentrated (beamed) to produce a radiation pattern of the desired form. Also, by *phasing*, or feeding two or more antenna elements in varying degrees of phase with respect to each other, the fields produced around each element can be made to reinforce or to cancel each other. The reinforcing or canceling is done in such a manner that the individual fields combine vectorially to produce a radiation field of maximum intensity along the desired paths in either or both horizontal and vertical planes.

Where a narrow and intense beam of energy is required, as in tracking radar applications, parabolic reflectors or combinations of reflector and director elements can be used to form the beam. The beam is formed in much the same manner that the reflector and lens in a bulls-eye lamp or focusing flashlight shape the beam of light (Fig. 4-23.)

LASER WELDER OPTICAL SYSTEM

Commercial lasers make much use of the optical devices discussed in this chapter. Figure 4-24 shows the block diagram of a laser welder optical system that incorporates many of the devices

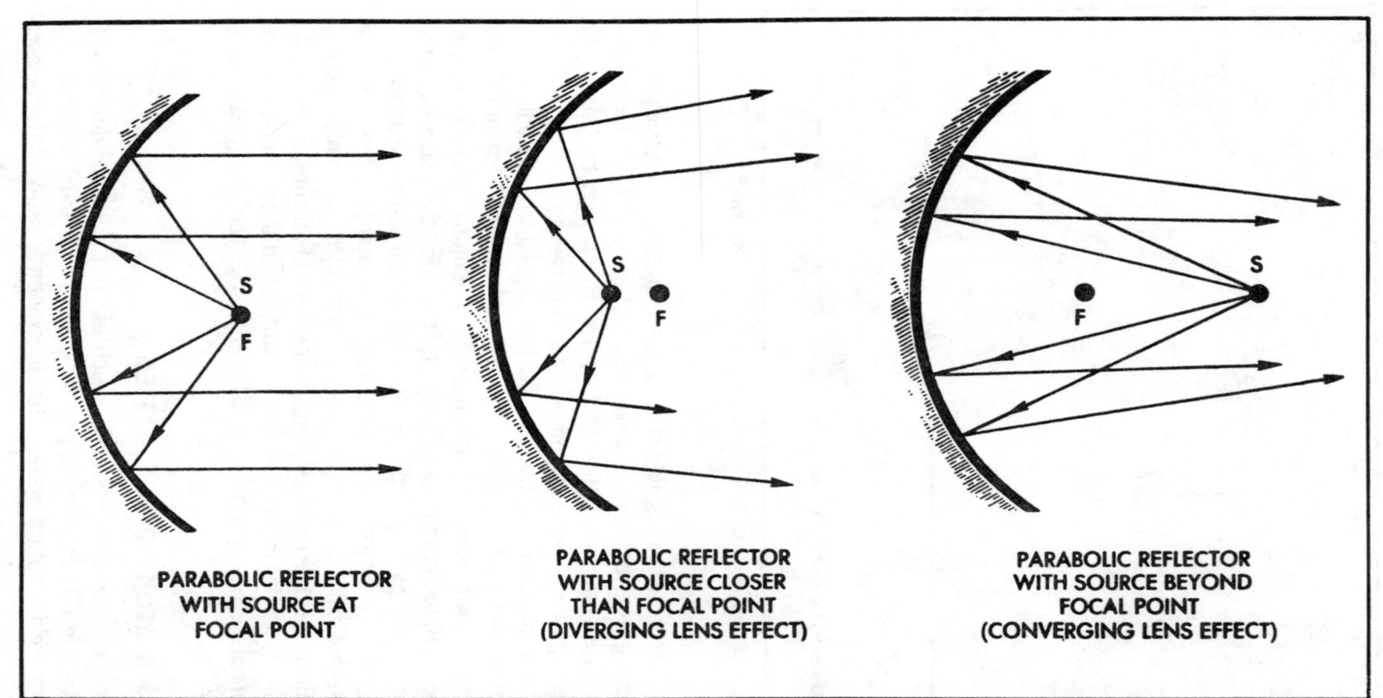

Fig. 4-23. Effects of parabolic reflector used in optics and electronics.

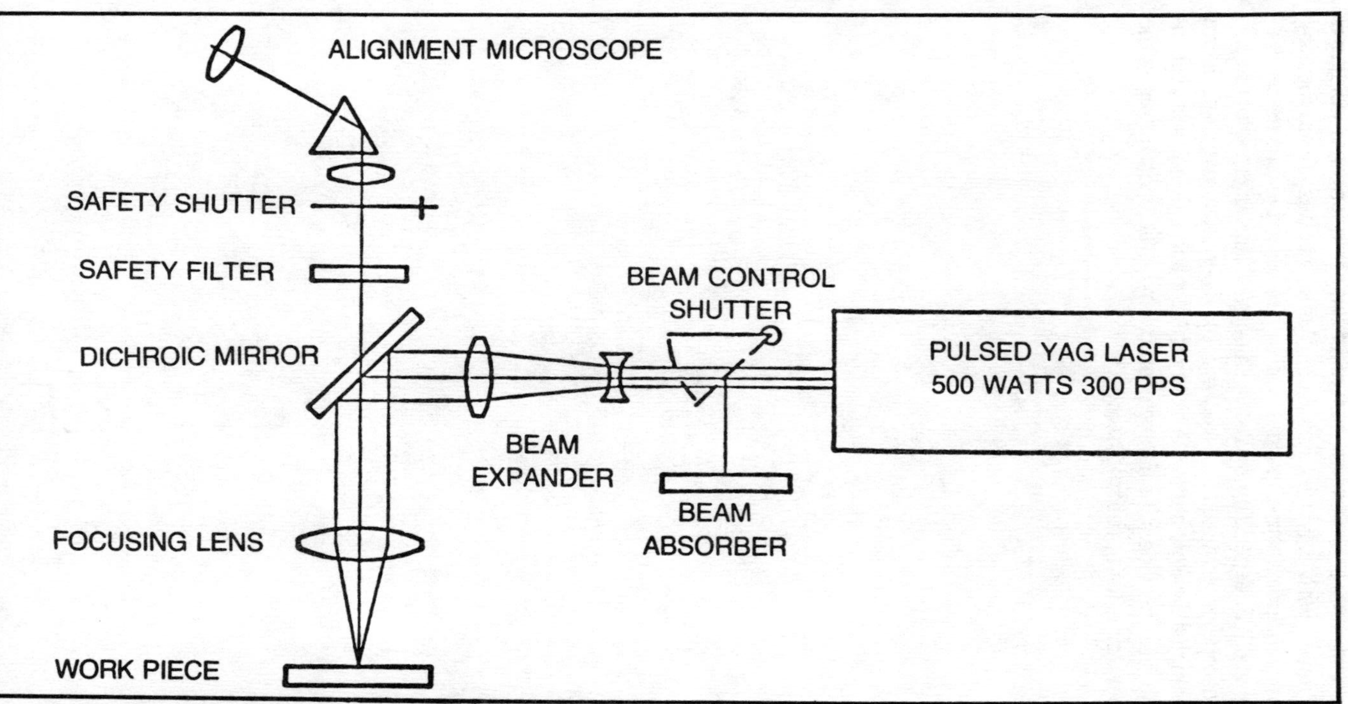

Fig. 4-24. Optical system of laser. (Courtesy of Raytheon Company).

discussed here. On-and-off control of the beam for seam welding is achieved with a fast external mechanical shutter. Beam-focusing optics include beam-expander and singlet objective lens systems of *f*/4 and *f*/2, depending on requirements. A binocular alignment microscope is provided for viewing coaxially with the laser beam through a dichroic mirror. A safety shutter and 1.06-micron absorption filter provide fail-safe eye protection for the operator when the operator is using the microscope.

Quantum Mechanics

A brief review of atomic theory is essential to understand the principles of operation of the laser. The treatment given here is only the essential knowledge.

THE ATOM

The atom is defined as the smallest particle of an element that retains all of the properties of the element. The atom is the smallest part of an element that enters into a chemical change, but it does so in the form of a charged particle. These charged particles are called *ions*, and they are of two types—positive and negative. A positive ion can be defined as an atom that has become positively charged. A negative ion can be defined as an atom that has become negatively charged. Ions of like charges tend to repel each other while ions of unlike charges attract, a fact well recognized from the study of magnetism and electricity (Chapter 3).

Atoms have been found to be divisible into more fundamental particles called electrons, protons, and neutrons. The electron was first discovered as the basic unit of electricity. It is a small, negatively charged particle much lighter than an atom. The proton is a positively charged particle with the same magnitude of charge as the electron but is much larger than the electron. The neutron is formed when a proton and electron are combined, and is neutral in charge.

The nucleus of the atom consists of a group of positive and neutral particles (protons and neutrons), surrounded by one or more

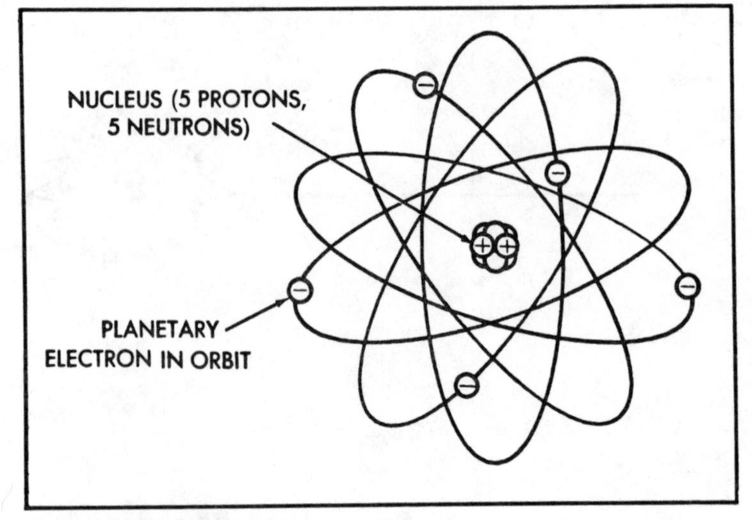

Fig. 5-1. Boron atom.

negative orbital electrons. Figure 5-1 shows the arrangement of these particles for an atom of the element boron. This concept of the atom can be likened to the solar system in which the sun is the central body around which the planets revolve in orbits at various distances from it. The electrons whirl about the nucleus of the atom much as the planets whirl about the sun.

In the lighter elements, the nucleus contains approximately one neutron for each proton while in heavier elements the neutrons will outnumber the protons. The nucleus of the helium atom consists of two neutrons and two protons. The mercury atom, a heavy element, has 80 protons and 120 neutrons in its nucleus.

PLANETARY ELECTRONS

Surrounding the positive nucleus of the atom is a negatively charged cloud made up of planetary electrons. Each of these electrons contains one unit of negative electricity equal in charge to the unit of positive electricity contained in the proton. In a normal atom, the number of planetary electrons is exactly equal to the number of protons in the nucleus. Therefore, the net charge of an atom is zero, since the equal and opposite effects of the positive and negative charges balance one another.

If an external force is applied to an atom, one or more of the outermost electrons can be removed. This is possible because the farther the electrons are from the nucleus, the less attraction they have to the nucleus. When atoms combine to form an elemental substance, the outer electrons of one atom will interact with the outer electrons of neighboring atoms to form bonds between the atoms. When bonding occurs in some substances, each atom retains its full complement of electrons. In other substances, one or more outer electrons will be gained or lost as a result of binding. The electron configuration of the atom is of great importance as the chemical and electrical properties of a material depend almost wholly upon the electron arrangement within its atoms.

The nucleus of the atom is well shielded by the electron cloud and does not enter into chemical or electrical processes. To dissipate the nucleus of an atom requires a vast amount of energy such as is released by each atom in the explosion of an atomic bomb.

THE HYDROGEN ATOM

The simplest of all atoms is that of the element hydrogen. The hydrogen atom is composed of a nucleus containing one proton and a single planetary electron. According to a concept developed by Niels Bohr (Fig. 5-2), the electron travels about the nucleus in a circular orbit having a fixed radius. As the electron revolves around the nucleus it is held in this orbit by two counteracting forces. One of these forces is called *centrifugal force*, and is the force which tends to cause the electron to fly outward as it travels around its circular

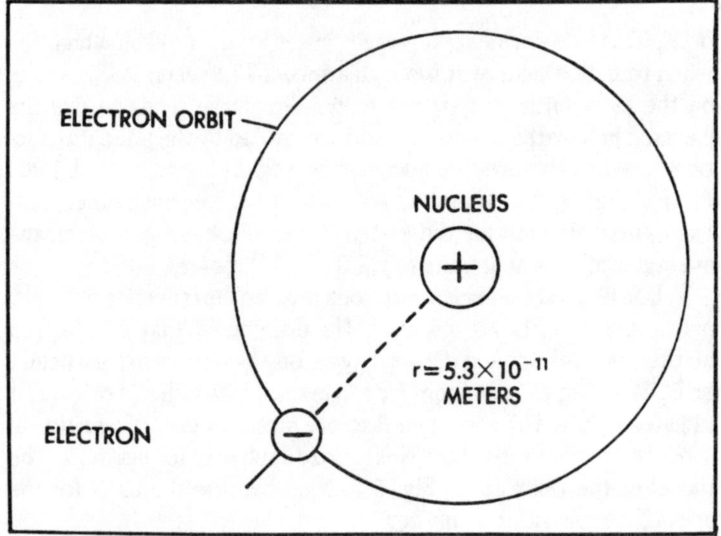

Fig. 5-2. Hydrogen atom.

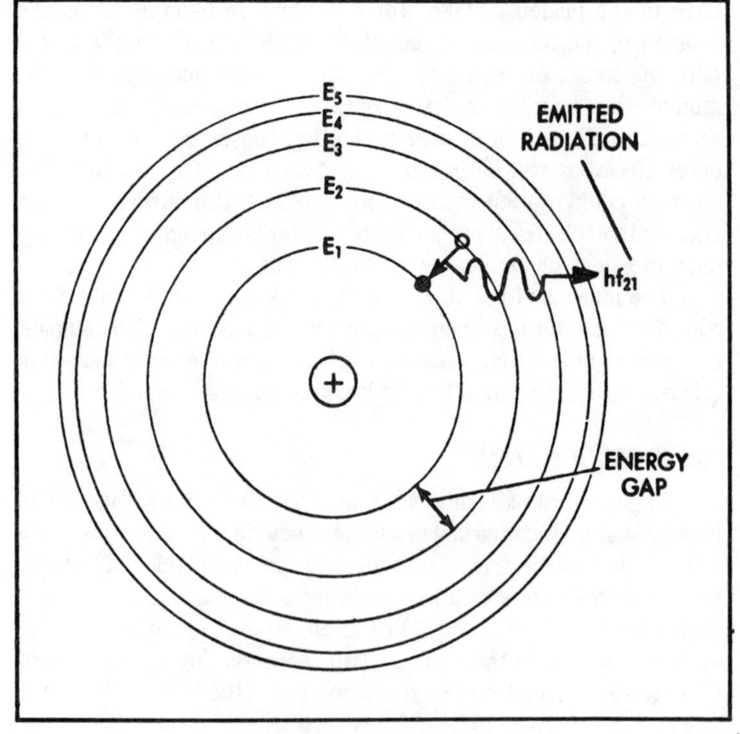

Fig. 5-3. Bohr model of hydrogen atom.

orbit. This is the same force which causes a car to roll off a highway when rounding a curve at too high a speed. The second force acting on the electron is *centripetal force*. This force tends to pull the electron in towards the nucleus and is provided by the mutual attraction between the positive nucleus and negative electron. At some given radius r, the two forces will exactly balance each other, providing a stable path for the electron. For the hydrogen atom, the average radius is approximately 5.3×10^{-11} meter.

Bohr stated that since electrons favored certain energy levels, only certain orbits are possible. He maintained that an electron neither radiates nor absorbs energy as long as it stays in a particular orbit. When an electron goes from one orbit to another, however, it radiates or absorbs energy in discrete amounts called *quanta*. In 1922 Bohr was awarded the Nobel prize in physics for his work. The model of the Bohr atom, Fig. 5-3, soon became the basis for the scientific explanation of matter.

QUANTUM PHYSICS CONCEPTS

The laser is often referred to as a *quantum electronic* device, thus necessitating the introduction of some important concepts of modern or quantum physics which directly apply to the laser. Certain pertinent facts relating to laser operation are presented here in expanded form as a foundation for the advanced topics treated in succeeding paragraphs. Max Planck was an obscure professor when he proposed the theory that if molecules of a material are excited, they will vibrate or oscillate and emit multiples of a unit amount of energy. This unit amount he designated as a quantum and defined it by a simple formula $E = hf$, where E is the amount of energy (the quantum), h is constant (Planck's constant) equal to 6.62×10^{-27} erg-sec, and f is the frequency of vibration or oscillation produced by the molecule as a tiny oscillator that can generate E, $2E$, $3E$, $4E$, etc. amounts of energy, but never a fractional amount of a quantum.

Planck's hypothesis that the energies of the electron oscillators responsible for radiation were quantized (restricted in certain integral multiples of a constant which now bears his name) was consistent with experimental findings. Planck's quantum hypothesis was successfully employed by Albert Einstein in explaining, among other phenomena, the photoelectric effect. In explaining the photoelectric effect, Einstein extended the quantum concept to the radiation itself. He assumed that light interacts with electrons in a metal as if the light was itself composed of discrete bundles of energy. Consequently, a light beam can be thought of as a stream of massless particles called photons which travel at the speed of light. Each particle contains the energy of hf joules.

ENERGY LEVELS

Since the electron in the hydrogen atom has both mass and motion, it contains two types of energy. By virtue of its motion the electron contains kinetic energy. Because of its position, it also contains potential energy. The total energy contained by the electron (kinetic plus potential) is the factor which determines the radius of the electron orbit. Orbit E_1 shown in Fig. 5-3 is the smallest possible orbit the hydrogen electron can have. For the electron to remain in this orbit, it must neither gain nor lose energy.

Light energy exists in tiny packets or bundles of energy called photons. Each photon contains a definite amount of energy depending on the color (wavelength) of light it represents. Should a photon of sufficient energy collide with the orbital hydrogen electron, the electron will absorb the photon's energy. The electron, which now has a greater-than-normal amount of energy, will jump to a new orbit farther from the nucleus. The first new orbit to which the electron can jump has a radius four times as large as the radius of the original orbit. Had the electron received a greater amount of energy, the next possible orbit to which it could jump would have a radius nine times the original. Each orbit may be considered to represent one of a large number of energy levels that the electron can attain. It must be emphasized that the electron cannot just jump to any orbit. The electron will remain in its lowest orbit until a sufficient amount of energy is available, at which time the electron will accept the energy and jump to one of a series of permissible orbits. An electron cannot exist in the space between permissible orbits or energy levels. This indicates that the electron will not accept a photon of energy unless it contains enough energy to elevate the electron to one of the allowable energy levels. Heat energy and collisions with other particles can also cause the electron to jump orbits.

Once the electron has been elevated to any energy level higher than the lowest possible energy level, the atom is said to be in an excited state. The electron will not remain in this excited condition for more than a fraction of a second before it will radiate the excess energy and return to a lower energy orbit. To illustrate this principle, assume that a normal electron has just received a photon of energy sufficient to raise it from the first to the third energy level. In a short period of time, the electron can jump back to the first level, emitting a new photon identical to the one it received.

A second alternative would be for the electron to return to the lower level in two jumps—from the third to the second, and then from the second to the first. In this case the electron would emit two photons, one for each jump. Each of these photons would have less energy than the original photon which excited the electron and would represent a longer wavelength of light.

This principle is used in the fluorescent light where ultraviolet light photons, which are not visible to the human eye, bombard a phosphor coating on the inside of a glass tube. The phosphor electrons, in returning to their normal orbits, emit photons of light that are of a visible wavelength (longer wavelength). By using the proper chemicals for the phosphor coating, any color of light may be obtained. The coloring of the screen of a television picture tube is an example.

In 1913, Niels Bohr gave the quantum theory new stimulus by proposing a quantized model for the hydrogen atom. In Bohr's model, the hydrogen atom is pictured as a small, positively charged nucleus, orbited by an electron. Bohr postulated that the rotational energy of the orbiting electron can have only certain discrete values. These values define a set of stable electron orbits; that is, while an electron is rotating in a stable orbit, it does not emit radiation. A region which separates allowed energy levels is called *an energy gap*. The electron can change energy only in a jump in which it either absorbs or emits a photon. The conservation of energy for an electron jump can be written $E_m - E_n = hf_{m,n}$ where E_m and E_n are two allowed energy states for the orbiting electron, and $f_{m,n}$ is the frequency of the emitted radiation. Figure 5-3 shows this for energy levels corresponding to $m = 2$ and $n = 1$. Using Bohr's model, the frequencies of the emitted radiation could be calculated. These calculated frequencies were found to agree almost perfectly with the observed characteristic spectrum of hydrogen.

In the years following the introduction of Bohr's model, the quantum theory grew in significance and scope. Some of the features of the Bohr atom provide a useful introduction to the important concepts necessary to the description of laser operation.

A concrete picture of the Bohr model in terms of energy is provided by Fig. 5-4, where energy is plotted on the vertical scale. The horizontal lines represent the allowed energy levels, and the vertical connecting lines represent examples of electron jumps, or transitions. The wavelength of the emitted radiation is given along the transition lines in Angstrom units (1 angstrom = 10^{-10} meter). Energy level E denotes the lowest energy level, or ground state, for the atom. The other levels (E_2, E_3, etc.) represent excited states. Such a representation is called an *energy-level diagram*. The usefulness of these diagrams in atomic physics can be appreciated when one realizes that the energy-level diagram for a given atom is peculiar to that type of atom. The energy-level diagram for an atom is to the atomic physicist what the schematic is to the electronics technician, because such a diagram provides important information concerning atomic behavior.

One feature of an atom's behavior not included in an energy diagram is the lifetime and population of levels. If the atom at some given time is in an excited state, its electrons are at any level except the ground state. The atom does not remain at this level if disturbed, but spontaneously jumps to some other level, and emits radiation. There is always a tendency, following radiation, for an atom to return to its lowest energy, or ground state. Consequently, it radiates, then spontaneously makes transitions downward, radiating at each level

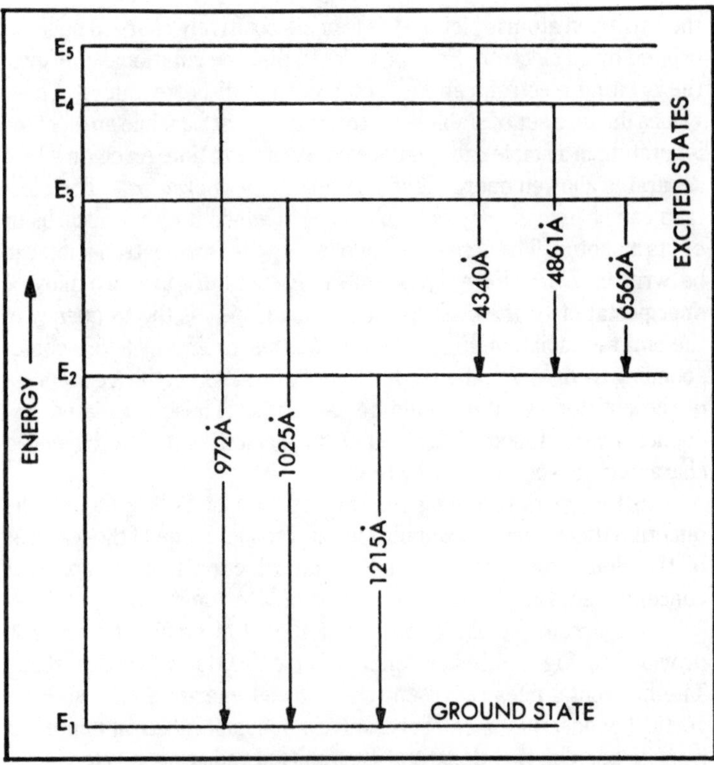

Fig. 5-4. Energy-level diagram for Bohr model of hydrogen atom.

until the ground state is reached. In a gas with unexcited atoms, almost all atoms are in the ground state.

The situation is complicated by the fact that the average time an atom remains in a state, before decaying to some lower level, depends on what state it is in to begin with. To complete the picture, a set of numbers must be made available which represents the mean lifetime of the electron in each of its possible states before it decays in spontaneous steps to the ground state. In general, these energy state lifetimes are quite short, less than 10^{-8} second. However, levels exist for which the lifetime is considerably longer. These levels are known as *metastable* states. Remember that each of the spontaneous transitions must conserve energy, therefore, each is accompanied by the emission of a photon. Photons emitted by spontaneous transitions are called spontaneous emissions. Since one atom does not know what another atom is emitting, there is a lack of cooperation among atoms, and the resulting emission produced is not coherent.

COMPLEX ATOMS

Although hydrogen is the simplest of all atoms, the basic principles just developed apply equally well to the atoms of more complex elements. The manner in which the orbits are established in an atom containing more than one electron is somewhat complicated and is part of the science of quantum mechanics. In an atom containing two or more electrons, the electrons interact with each other and the exact path of any one electron is very difficult to predict. Each electron will lie in a specific energy band and the orbits can be considered as an average of the electron's position.

SHELLS AND SUBSHELLS

The difference between the atoms, insofar as their chemical activity and stability is concerned, depends upon the number and position of the particles included within the atom. Atoms range from the simplest, the hydrogen atom containing one proton and one electron, to the very complex atomic structures such as silver, containing 47 protons and 47 electrons. In general, the electrons within the atom reside in groups of orbits called *shells*. These shells are elliptically shaped and are assumed to be located at fixed intervals as predicted by the Bohr concept. Thus, the shells are arranged in steps that correspond to fixed energy levels. The shells, and the number of electrons required to fill them, can be predicted using ***Pauli's exclusion principle***. Simply stated, this principle specifies that each shell will contain a maximum of $2n^2$ electrons, where n corresponds to the shell number starting with the one closest to the nucleus. By this principle the second shell, for example, would contain $2(2)^2$ or 8 electrons when full.

In addition to being numbered, the shells are also given letter designations as pictured in Fig. 5-5. Starting with the shell closest to the nucleus and progressing outward, the shells are labeled *K, L, M, N, O, P,* and *Q*, respectively. The shells are considered to be full or complete when they contain the following quantities of electrons: 2 in the *K* shell, 8 in the *L* shell, 18 in the *M* shell, and so on, in accordance with the exclusion principle. Each of these shells is a major shell and can be divided into subshells of which there are four, labeled 1, 2, 3, and 4, respectively. Like the major shells, the subshells are also limited as to the number of electrons which they can contain. Subshell 1 is complete when it contains 2 electrons, subshell 2 when it contains 6, subshell 3 when it contains 10, and the last or subshell 4 when it contains 14 electrons.

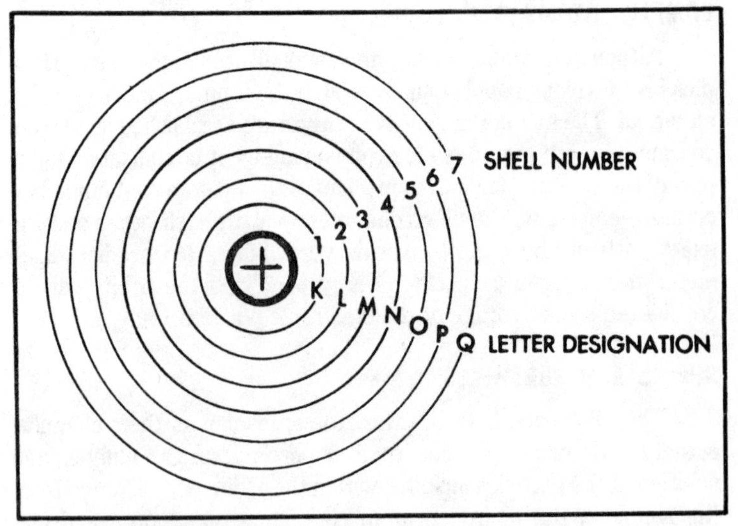

Fig. 5-5. Shell designations.

Inasmuch as the *K* shell can contain no more than two electrons, it must have only one subshell. The *M* shell is composed of three subshells: 1, 2, and 3. If the electrons in subshells 1, 2, and 3 are added, their total is 18, the exact number required to fill the *M* shell. This relationship exists between the shells and subshells up to and including the *N* shell. Beyond the *N* shell, the actual number of electrons required to fill a shell has not been experimentally determined.

To drive electrons out of the shells of an atom requires the internal energy of the atom to be raised. This raised energy may be obtained through bombardment by photons or by subjecting the atom to electronic fields. The amount of energy required to free electrons from an individual atom is called the ***ionization potential***.

The ionization potential necessary to free an electron from an inner shell is much greater than that required to free an electron from an outer shell. Also, more energy is required to remove an electron from a complete shell than from an unfilled shell.

The discussion of the atom, atomic structures, and energy levels is included because of the important role it has played in laser development. Laser operation requires (1) an active material that produces stimulated emission of radiation, (2) an excitation source that will pump power into the active material, and (3) a resonant structure.

STIMULATED EMISSION

Ordinary light sources do not rely solely on spontaneous emissions for their output. Spontaneous transitions are not the only means by which a particular atom returns to its ground state. Consider the hypothetical case of a coherent light (photon) beam traversing atoms of a gas. When the frequency of such a source coincides with one of the frequencies of spontaneous emission, atoms are induced to make transitions between two particular energy levels whose difference is ΔE, thus satisfying the relationship $\Delta E = hf$. An important result is that transitions from the upper to the lower energy states are induced, in addition to those in the opposite direction. The probability that the induced transition will be in one direction rather than the other depends only on which level the majority of atoms are located. When the transition is from a lower to an upper energy level (accompanied by the loss of one photon from the beam), it is called *absorption*. When the transition is from upper to lower energy levels a photon is emitted, and it is called *induced* or *stimulated emission*.

If equal numbers of atoms are in two levels, the beam intensity remains constant in traversing the gas. With more atoms in the upper than in the lower state, the beam sees a net gain of photons, or is amplified by stimulated emissions in traversing the gas. Since photon emissions are induced by interaction with the photons in the beam, which are assumed to be coherent, they are all in-phase, and the amplified beam is coherent, as shown in Fig. 5-6. In this way, a single induced emission is amplified into an intense coherent beam. The remaining problem is to obtain the higher concentration of atoms in the upper energy level necessary for amplification. This is known as *population inversion*.

Stimulated emission is the basis of laser operation. When the photon is absorbed by the atom, the energy of the photon is converted to internal energy of the atom. The atom's electrons are raised to a "high energy" or "excited" state and later radiate this energy spontaneously, emitting a photon and returning to the ground state. Phosphorus illustrates this principle; it glows when hit by light and continues to glow for a period of time after the light is removed. Using ultraviolet light makes it glow as long as the ultraviolet light is present; however, as the energy source is increased, the glowing state ceases. The light energy is released in-phase (polarized), but it is not lasting action even though this light is monochromatic (one frequency). This phenomenon is the beginning of the laser principle,

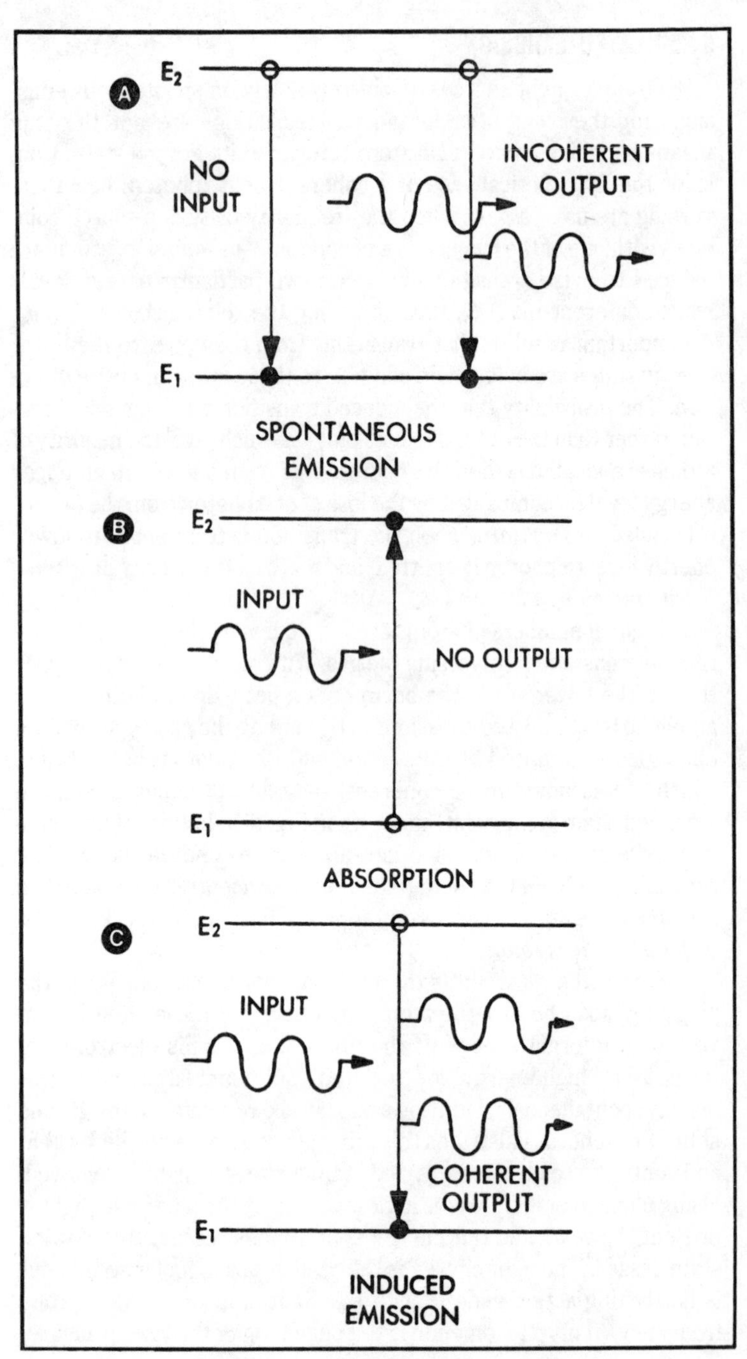

Fig. 5-6. Atomic transmissions producing emission and absorption.

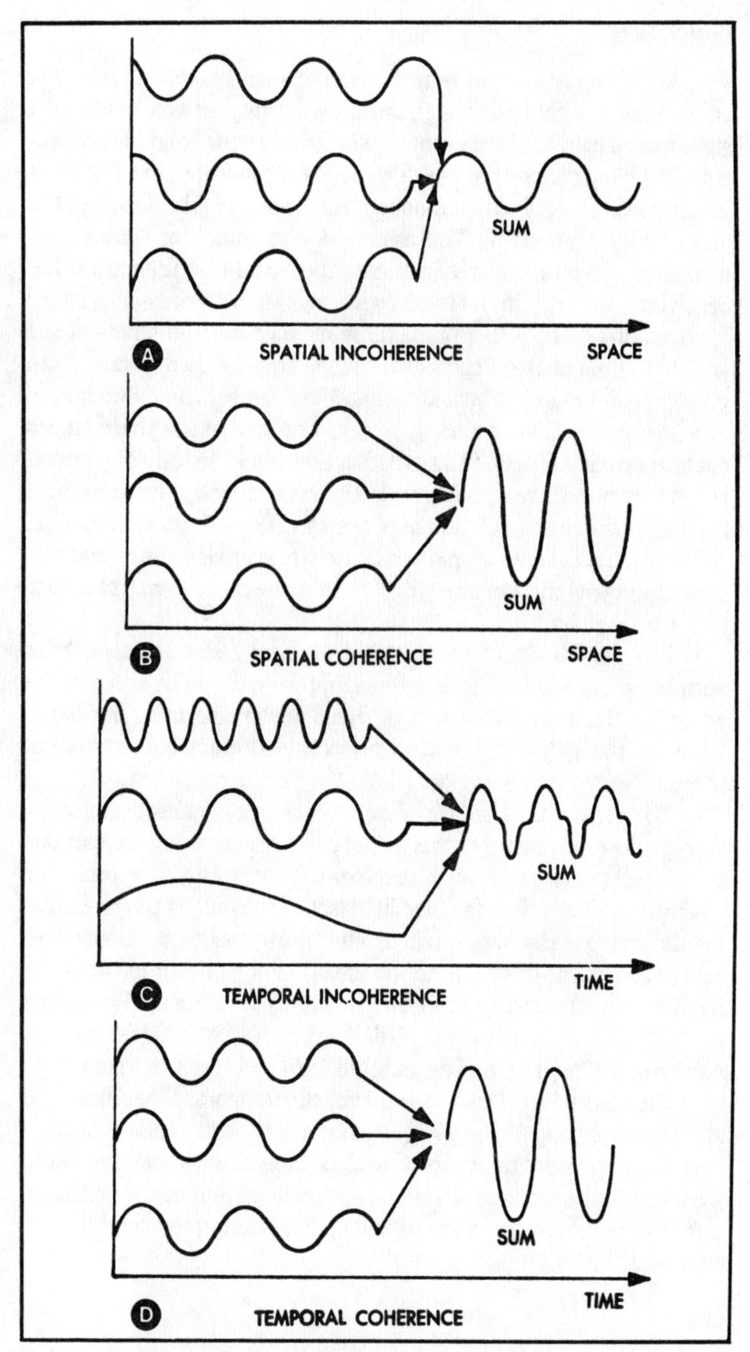

Fig. 5-7. Spatial and temporal coherence effects and waveforms.

COHERENCE

You may recall from your study of electronics that an effective transmitter is a generator of electromagnetic waves which radiates a significant amount of coherent power in a narrow band of frequencies, including the one desired. The need for coherence in an efficient generator is sometimes overlooked. An example is the design of the antenna for a receiver. The assumption is sometimes made that increasing the antenna area increases the signal-to-noise ratio at the receiver detector. This is true only if the phase of the incoming signal is constant or varies in a predictable manner over all points of the antenna. If the phase of the signal changes in a random manner from point to point over the antenna, then the detector can only sum the absolute value of the incoming power. The sum of the phase angles could approximate zero and all modulation would be lost. This correlation of phase in the signal is precisely what is meant by *coherence*. An instantaneous correlation of phase from point to point in space is called *spatial coherence*, and a consistent correlation in-phase at two neighboring points over a length of time is called *temporal coherence* (refer to Fig. 5-7).

The generators of radio, television, and radar signals exhibit both temporal and spatial coherence in the emitted signal. Until the advent of the laser no sources of signal power operating in what is known as the *optical spectrum* were coherent in any but a statistical sense.

The reason for the lack of coherence in optical sources other than the laser is related to the lack of correlation among the motions of the electrons, since each electron behaves as a tiny oscillator which emits light. For familiar light sources, such as tungsten filaments and gas-discharge tubes, electrical energy is supplied to create conditions favorable to the emission of light; for example, by heating the filament or exciting the atoms in the gas. The actual emission process is not controlled. Each oscillator radiates independently of its neighbors. The emitted light, which is the sum of all individual radiations, lacks both spatial and temporal coherence. The idea was first proposed in 1958 to maintain a constant phase relation over the oscillators by stimulating their emission with a wave of the frequency to be radiated. Two years later, using stimulated emission, pulses of coherent optical radiation from a single ruby crystal were achieved.

6

Microwaves and Resonant Cavities

The first stimulated-radiation devices were masers. As mentioned, the letters of the acronym, maser, stand for microwave amplification by stimulated emission of radiation. When optical stimulated-radiation devices were invented, they were referred to as optical masers. Now, of course, they are called lasers. This chapter considers principles common to both masers and lasers.

Many of the devices used in ordinary microwave systems are also used in maser systems. Klystrons are used to produce pumping energy for masers. Sometimes the maser action takes place in a traveling-wave-tube structure. Microwave energy is conducted in and out of a maser by waveguides. The maser action is intensified by the use of a resonant cavity, which is a section of waveguide with the ends closed off.

Resonant cavities are also used with optical masers, or lasers. Most lasers have parallel reflecting surfaces at the ends of the cavity. Since a single-wavelength resonator at optical wavelengths would be much too small to be practical, a special resonator is built with dimensions that are thousands of times greater than a single wavelength. The reflecting surfaces, or mirrors, are installed facing each other in the cavity. The mirrors must be positioned just right. They must be parallel and they must be a certain number of wavelengths apart. One of the surfaces is completely silvered, as an ordinary mirror; the other is only partially silvered and remains somewhat transparent, like the lenses in mirrored sunglasses. Exactly how the mirrors are used in lasers is covered in later chapters. Here we are concerned with the general principles of cavities.

WAVEGUIDES

Basically the waveguide is a special class of transmission line. The development of the waveguide from the two-wire transmission line is illustrated in Fig. 6-1. In a two-wire line, a quarter-wave section at the signal frequency presents a highly resistive impedance. Thus, a quarter-wave shorting stub can be used as an efficient insulator. If the stub offers adequate mechanical strength, it can be used to support the line while insulating it.

A rigid quarter-wave stub connected between the leads of a two-wire open transmission line as shown in Fig. 6-1A provides a starting point from which to discuss the development of a waveguide. In this type of transmission line, efficient operation is limited to a single frequency. For all other frequencies, the insulating properties of the stub are decreased. A given stub represents a quarter wavelength only at one frequency, usually called the *resonant* frequency. At signal frequencies higher than resonance, the stub appears longer than a quarter wavelength and represents an inductive impedance. At frequencies below resonance, the stub appears shorter than a quarter wavelength and offers a capacitive impedance. The greater the difference between the signal frequency and the resonant frequency of the stub, the smaller the total impedance offered by the stub. Since the stub normally has a very high Q, the effective useful range of frequencies is critical.

A metallic quarter-wave stub insulator can be placed at any point along a two-wire line. Use of several such stubs, placed along the line as shown in Fig. 6-1B, improves the mechanical support of the line. Whether the stubs are used to support the line from above or below makes no difference.

If the stubs are connected together, the conductivity of the line is improved. Figure 6-1C shows adjacent stubs connected by a switch. When the switch is open, each stub is excited separately by the signal in the main line. Standing waves are present on the line. Stub 1 is excited first, stub 2 at some time later. When the switch is closed, however, stub 2 is excited partially by the signal on the main line and partially by stub 1 (through the switch). If the switch is connected to corresponding points on the two stubs, the relative phase relationship of the signal at the connections is the same. Thus,

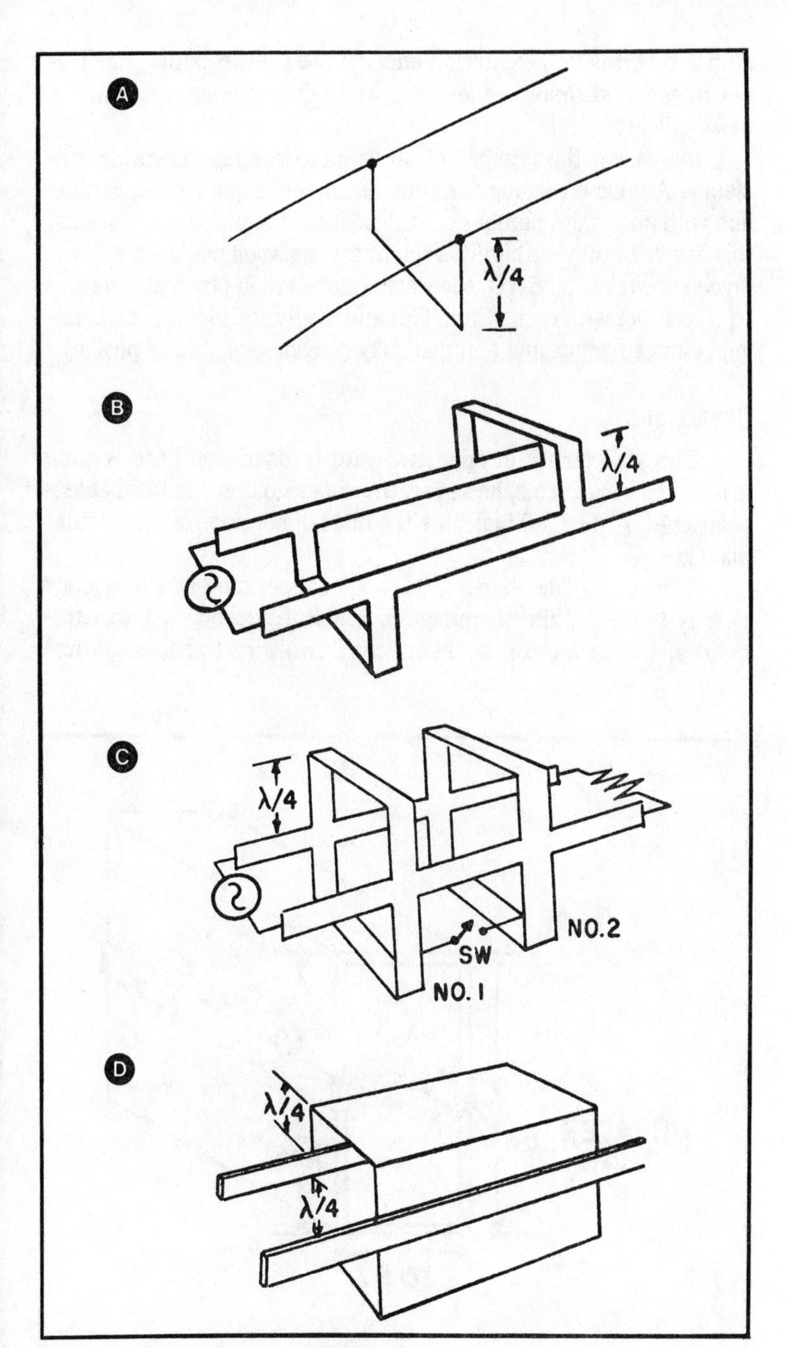

Fig. 6-1. Development of a waveguide.

stub 2 requires less excitation energy; the parallel paths offer less resistance (and therefore less copper loss), and energy transfer is more efficient.

Increasing the number of stubs increases the operating efficiency. A hollow rectangular tube effectively represents a parallel line with an infinite number of stubs. Given the proper dimensions, this device becomes a highly efficient transmission line. In the waveguide shown in Fig. 6-1D, successive stubs are in physical and electrical contact with each other. This effectively connects corresponding points together and automatically provides for proper phasing.

Dimensions

The stub length in open two-wire transmission lines is quite critical. In waveguides, however, the dimensions are somewhat less restricted. Figure 6-2 indicates the main dimensions of rectangular waveguides.

The waveguide shown in Fig. 6-2 is operating at a frequency that is the approximate minimum usable frequency. At this frequency, the waveguide is effectively a two-wire transmission line

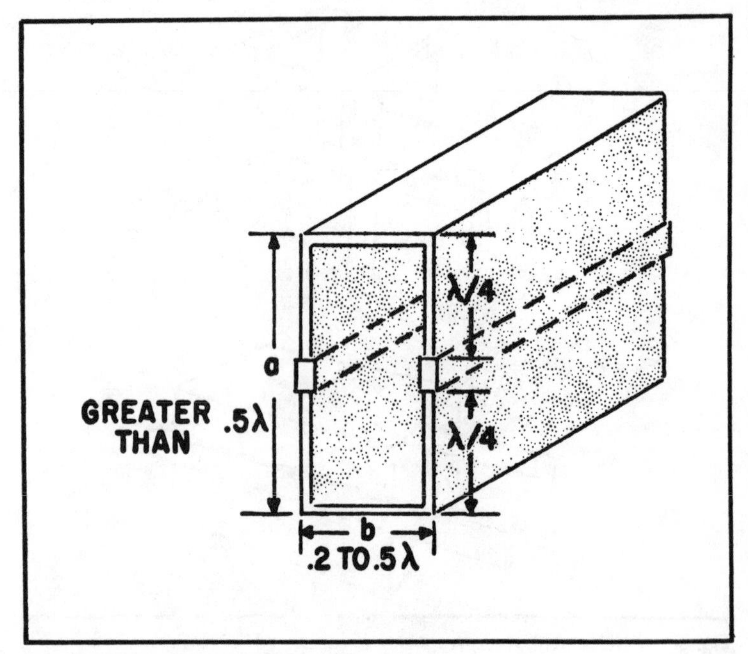

Fig. 6-2. Dimensions in a rectangular waveguide.

with an infinite number of shorted stubs (indicated by dashed lines in the figure). At any signal frequency higher than the one illustrated, the width of the two leads of the effective line is increased, and the remaining portion of the stub then represents a quarter wavelength. At frequencies lower than the one illustrated, however, the stubs are less than a quarter wave and the leads are not adequately insulated. Thus it can be assumed that dimension a of the waveguide is not critical, except that the signal must be at least a certain minimum frequency.

In open transmission lines, the spacing of the lines is critical only with respect to voltage breakdown. If the wires are placed too close together, arcing can occur and energy losses become excessive. The same principle applies to dimension b of the waveguide.

The main limitation on the length of either the open line or the waveguide is the energy losses incurred by the signal in transit through the device from one end to the other.

The frequency at which the quarter-wave stub sections appear to meet and cease to exist is called the ***cutoff frequency***, and the corresponding wavelength is called the ***cutoff wavelength***. Thus the analogy provides a basis for the formula $\lambda_{CO} = 2a$, where λ_{CO} is the wavelength at the cutoff frequency of the waveguide, and a is the wide dimension as indicated in Fig. 6-2. In practical waveguides, dimension b is normally between 0.2 and 0.5 wavelength at the cutoff frequency.

Energy in Waveguides

Energy is transferred within the waveguide by electromagnetic fields; currents and voltages merely aid in establishing these fields. In understanding the operation of waveguides, it is important to know when a good current path is required and where the voltage must be high. It is also important to know where strong fields will exist and to understand the composition of the fields. It is normally by means of fields that energy is introduced into or removed from the waveguide. The operation of the waveguide involves two field components—the electric field and the magnetic field. These were discussed in general terms in Chapter 3.

Electric Field. As discussed earlier, the existence of an electric field indicates that a difference of potential exists between two points. An excited transmission line has a standing wave along its path; part of the line is positive and part is negative. An instantaneous field is set up between the wires. The standing wave, and therefore the electric field, varies in magnitude (intensity) progressively along

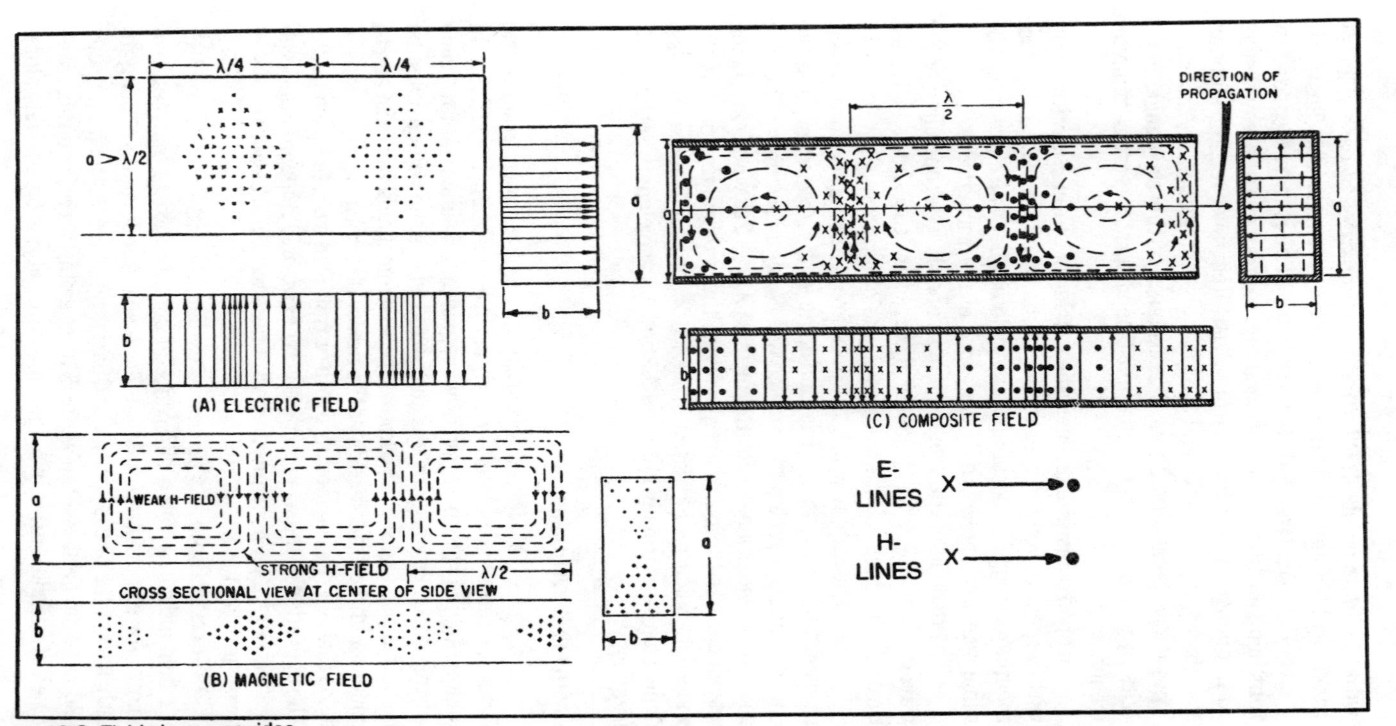

Fig. 6-3. Fields in waveguides.

the line. Since the exciting voltage varies at a sinewave rate, the electric field also varies sinusoidally.

The electric field, as it exists in a waveguide, is shown in Fig. 6-3A. This is an instantaneous representation of the field at its peak. Over a period of time (during each cycle), the field decreases to zero, reverses polarity and builds up to another maximum, decreases to zero, then reverses and expands again. Successive zero points occur at half-wave intervals along the length of the line; successive maxima, alternating in polarity, also occur at half-wave intervals.

Note that in illustrating the direction of the fields, standard practice is to use dots (.) and crosses (x). The dot indicates that the direction of the field is toward the observer. The cross indicates that the field is away from the observer.

Also observe that the lines of force that make up the electric field, the E-lines, exist between dimension b walls and are perpendicular to dimension a (wide) walls. The field has maximum intensity halfway between dimension b walls, decreasing toward zero as it nears a wall. The importance of this will be clear shortly.

Magnetic Field. Any expanding and collapsing electric field creates an alternating magnetic field. A visual representation of the sinusoidal magnetic field at an instantaneous maximum point is shown in Fig. 6-3B. Note that the field forms closed loops approximately half a wavelength long, with successive loops reversed in polarity. Note also that the magnetic lines, or H-lines, are parallel to the walls of the waveguides.

The magnetic field is strongest at the edges of the waveguide and weakest at the center of each loop (where the standing wave of current is zero at all times).

Composite Field. Electromagnetic waves are made up of electric and magnetic field components. Energy travels down the waveguide in the form of wavefronts.

As a wavefront progresses down a waveguide, it interacts with a wavefront reflected from the other end. This interaction results in standing waves similar to the standing waves of current and voltage on transmission lines. Figure 6-3C illustrates the composite electromagnetic standing waves in a section of waveguide.

In the figure, note that the electric and magnetic components of the composite field are oriented perpendicular to each other and to the direction of propagation. Also note that the points of maximum electric and magnetic field intensity coincide, as do the points of minimum field strength. Propagation of energy in waveguides differs in some respects from that in free space.

In free-space propagation, the transmitted wave expands in an ever-widening circle. In waveguides, it is confined within the conductive walls, and no electromagnetic field exists outside the waveguide. Instead, as the wave travels down the guide, it strikes the walls at an angle and is reflected at an equal angle. If energy propagation is to be sustained within the waveguide, two major *boundary* conditions must be fulfilled.

Boundary Conditions. The first boundary condition that must be met is that there must be no component of the electric field tangent to the walls of the waveguide. An electric field is equivalent to a voltage; when impressed across a perfect conductor (the walls of a waveguide, for instance), it is decreased effectively to zero. Thus the parallel component of the E-field diminishes to zero as it approaches the walls. The remaining component of the E-field is perpendicular to the walls in dimension a.

The second boundary condition concerns the magnetic field. No component of the magnetic field can exist perpendicular to the inner surface of the waveguide. Since the electric and magnetic components of the composite field are mutually perpendicular, satisfaction of one boundary condition usually guarantees satisfaction of the other as well.

Circular Waveguides

Waveguides have been developed which have a circular cross section as well as those having a rectangular cross section. Although in less common usage than rectangular waveguides, circular ones have characteristics that make them particularly useful in some applications. In general, the development of the circular waveguide can be explained in the same manner as that of the rectangular.

The two conductors represented by the dashed lines in Fig. 6-4 are assumed to be part of the waveguide wall. The remaining part of the wall represents the quarter-wave stub sections, which act as the insulators.

The cutoff frequency of the circular waveguide is 1.71 times the diameter of the guide. The diameter must be 1.17 times dimension a of a rectangular waveguide having the same cutoff frequency.

Modes in Waveguides

As mentioned previously, at frequencies higher than cutoff, the wide dimension of the waveguide becomes increasingly greater than

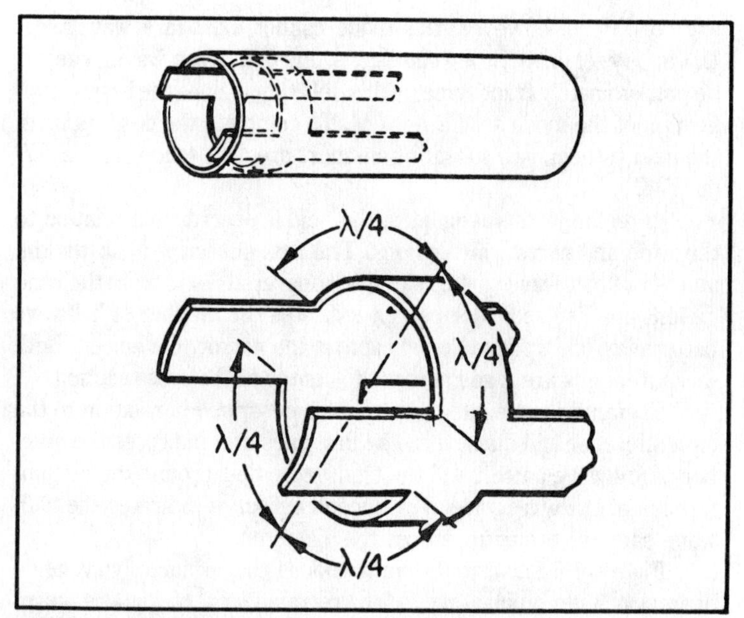

Fig. 6-4. Circular waveguide.

a half wavelength. At a frequency twice that of cutoff, this dimension is a full wavelength, with each stub representing a halfwave section. If the waveguide and its fields are to fulfill the boundary conditions required for propagation, the field configuration must be quite different from that previously indicated. In other words, the waveguide must assume a different mode of operation.

The normal configuration of the electromagnetic field in a given waveguide is called the ***dominant mode of operation*** of that waveguide. Other modes are also possible in any waveguide, and some of these are in common use. Resonant cavities also have modes of operation. One called the TE_{00} mode is often used in lasers and the power output of lasers is often specified for this mode.

Any field configuration can be classed as either a ***transverse electric*** mode or as a ***transverse magnetic*** mode, abbreviated *TE* and *TM*. In a *TE* mode, all parts of the electric field are transverse (perpendicular to the length of the waveguide), and no *E*-lines are sustained parallel to the direction of propagation. In a *TM* mode, the plane of the *H*-field is transverse, and no *H*-lines are parallel to the direction of propagation.

A wavefront in free space is in a *TEM* mode, since both fields are perpendicular to the direction of propagation. Due to the boundary conditions, however, this mode cannot exist in a waveguide. Under any set of operating conditions, one field in any waveguide will be predominately transverse—this field determines the basic classification of the mode as *TE* or *TM*. To complete the description of the field pattern, two subscript numbers are used following the *TE* or *TM*.

In rectangular waveguides, the field is described in relation to the wide and narrow dimensions. The first subscript indicates the number of half-wave patterns of the transverse field across the wide dimension. The second subscript indicates the number of halfwave patterns of the transverse field across the narrow dimension. Both measurements are made across the center of the cross section.

In circular waveguides, the field is described in relation to the circumference and diameter. The first subscript indicates the number of full-wave patterns of the transverse field around the circumference of the waveguide. The second subscript indicates the half-wave patterns occurring across the diameter.

Figure 6-5 illustrates several modes that occur in waveguide operation. The dominant mode for a rectangular waveguide is shown in Fig. 6-5C, and the dominant mode for a circular waveguide is shown in Fig. 6-5F. Note that in Figs. 6-5D and 6-5E the electric field is axial; therefore, the magnetic field is transverse, and the modes are designated *TM*.

Excitation of Waveguides

As with any transmission line, the waveguide must be properly excited or fed in order to carry the energy from one point to another. Waveguides, however, do not have the two convenient connection points possessed by the conventional rf line. In waveguides, the energy is in the form of an electromagnetic wave; therefore, the excitation device must cause these fields to be created within the guide. In masers, the excitation device is actually within the guide. In conventional microwave systems, other excitation techniques are used. These same techniques are also used in the extraction of energy from the waveguide—in both masers and conventional microwaves. The techniques use electric fields, magnetic fields, or electromagnetic fields.

Electric Field. Inserting a small probe or antenna into a waveguide and applying an rf signal causes current to flow in the probe. This current sets up an electrostatic field; *E*-lines detach themselves from the probe and propagate in the waveguide. The propagating

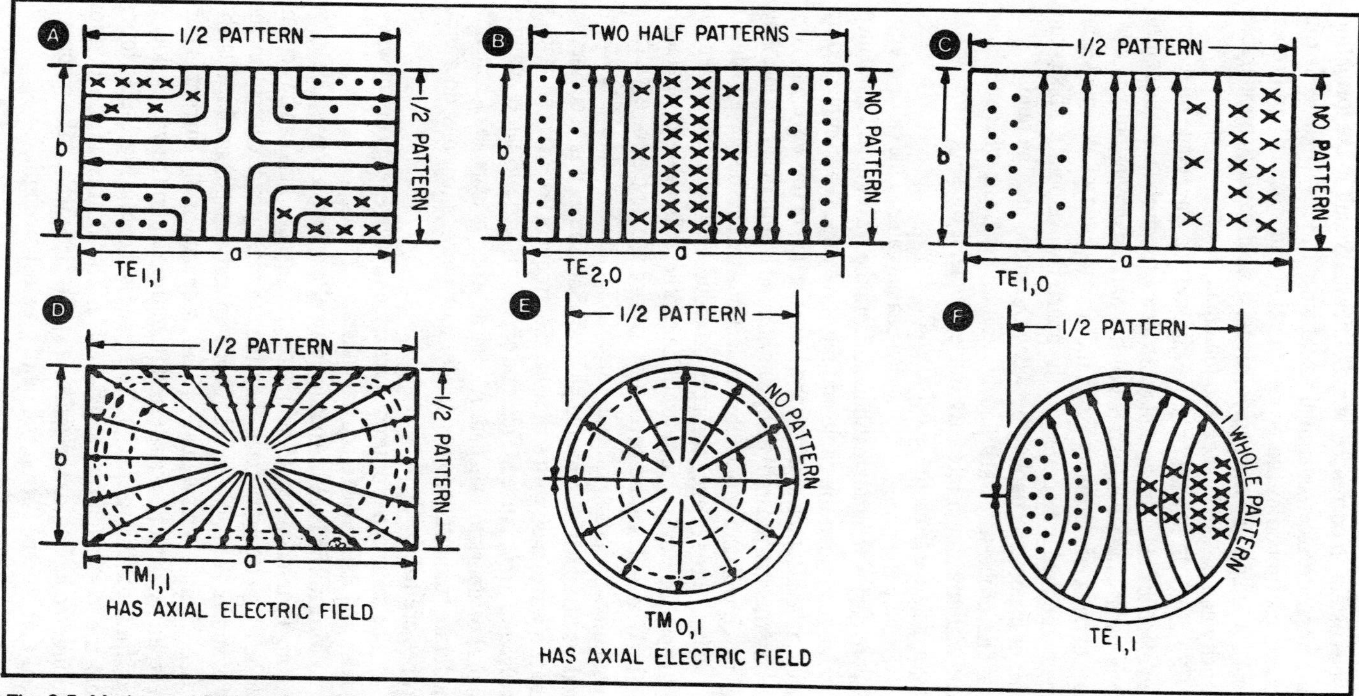

Fig. 6-5. Modes and field configurations in waveguides.

E-lines establish corresponding *H*-lines and result in the formation of a composite electromagnetic field within the waveguide. If the probe is positioned in the proper location and supplied the proper signal for a specific waveguide mode, a field having considerable intensity can be established and maintained.

Figure 6-6A illustrates the excitation of a waveguide by means of an electric field. The best place to locate the probe is in the center of the wide dimension, parallel to the narrow dimension, and one-quarter wavelength away from the shorted end of the guide, as shown (the field is strongest at the quarter-wave point). This is the point of maximum coupling between the probe and the field.

The probe will work equally well at any point where *E*-lines exist with maximum intensity (for example, a three-quarter wave-length distance from the shorted end).

Usually the probe is fed with a coaxial cable. This cable is kept short to derive the greatest benefit from the waveguide. Impedance matching between the cable and the waveguide is accomplished by varying the distance of the probe from the end of the waveguide (by moving the shorted end) or by varying the length of the probe. Any mismatch will cause unwanted reflections within the waveguide.

The degree of excitation can be reduced by reducing the length of the probe, moving it out from the center of the *E*-field, or shielding it. Where it will be necessary to vary the degree of excitation frequently, the probe is made retractable, and the end of the wave-guide is fitted with a movable plunger. Sometimes it is necessary to excite a waveguide with a wide band of frequencies. In these cases a ***wideband probe*** is used. This kind of probe is large in diameter and is conical or doorknob shaped. A conical probe is capable of handling moderate amounts of power, but for high-power situations, doorknob probes are used.

The same types of probes are used to take energy out of wave-guides and deliver it to coaxial cables.

Magnetic Field. Another way of exciting a waveguide is by setting up a magnetic *H*-field in the waveguide. This can be accomplished by placing a small loop that carries a high current in the waveguide. A magnetic field builds up and expands until it fills the space within the waveguide. If the frequency of the current is correct, energy will be transferred from the loop to the waveguide. A loop for transferring energy into a guide is shown in Fig. 6-6B. Notice that the loop is fed by a coaxial cable. The location of the loop for optimum coupling to the guide is at the place where the magnetic field that is to be set up will be of greatest strength.

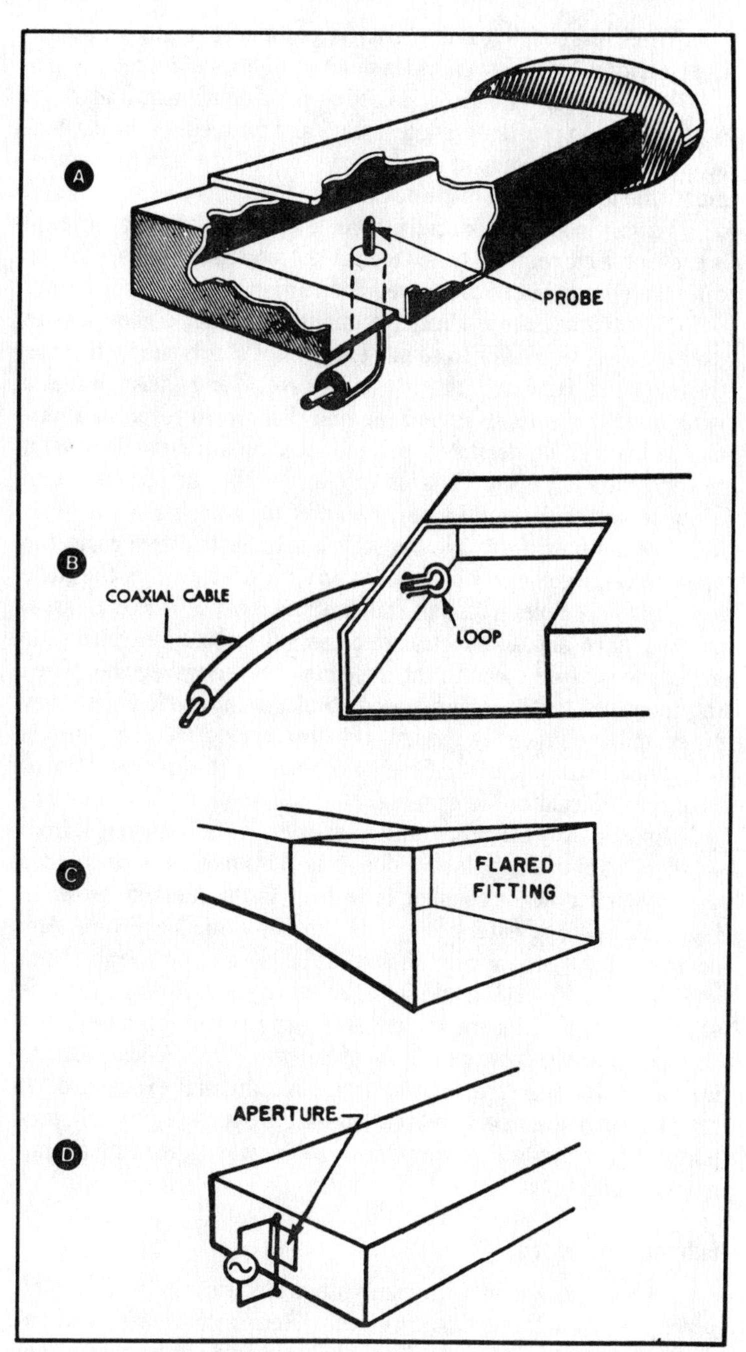

Fig. 6-6. Exciting a waveguide.

When less coupling is desired, the loop can be rotated or repositioned until it encircles a smaller number of lines of force.

When an excitation loop is used in some equipment (radar), its proper location is often predetermined and fixed either during construction or final tuning at the factory. In test or laboratory equipment, the loop is often made adjustable.

When a loop is introduced in a guide at a point where an H-field is present, a current will be induced in the loop itself. Thus, the loop can take energy out of the waveguide as well as put energy into it.

Electromagnetic Field. It might seem that a good way to excite the waveguide or to couple energy out of it is simply to leave the end open. However, this is not the case. When energy leaves a waveguide, fields exist around the end that would result in a mismatch. In other words, reflections and standing waves would result if the end were left open. Thus, simply leaving the end open is not an efficient way of getting the energy out of the waveguide.

For the energy to move smoothly into or out of a waveguide, the opening of the guide may be flared like a funnel. This makes the guide similar to a V-type antenna. This method, shown in Fig. 6-6C, is called a *horn* and, in effect, eliminates reflection by matching the impedance of free space to the impedance of the waveguide. When the mouth of the horn is exposed to electromagnetic fields, they enter and are gradually shaped to fit the waveguide. The horn is directional in characteristic. It sends or receives the greatest amount of energy in front of the opening.

Another way for either putting energy into or removing it from waveguides is through slots or openings. This method is sometimes used when very loose coupling is desired. In this method, shown in Fig. 6-6D, energy enters the guide through a small aperture. Any device that will generate an E-field can be placed near the aperture, and the E-field will expand into the waveguide. A single wire is shown; on it, E-lines are set up parallel to the wire because of the voltage difference between parts of the wire. The E-lines expand first across the aperture, then across the interior of the waveguide. If the frequency is correct and the size of the aperture properly proportioned, energy will be transferred to the waveguide with a minimum of reflections.

Extraction of Energy

In waveguides, as with many other electrical networks, *reciprocity* exists in the excitation system. Reciprocity means that energy can be either introduced into the waveguide or removed from the waveguide by using the same coupling devices and methods, with the same efficiency of transfer.

RESONANT CAVITIES

By the simple expedient of closing the ends of a waveguide to form an enclosure, a cavity is formed. If this cavity is then excited in a proper manner by the proper frequency, it will resonate. The resonant frequency of the cavity is determined by the dimensions, which are somewhat more critical than those of a waveguide.

In order to resonate, the cavity makes use of the distributed inductance and capacitance of the enclosing box. This limits the practical use of resonant cavities to microwave frequencies and laser frequencies. Resonant cavities normally possess a value of Q considerably in excess of any value obtainable with lumped constants at lower frequencies.

Fields in Resonant Cavities

As in the waveguide, operation of the resonant cavity is best described in terms of the fields rather than in terms of currents and voltages. Cavities can have various physical shapes. A chamber enclosed within conductive walls can resonate at any of several frequencies, thus produce a number of field configurations, depending on the size and shape of the chamber. As in waveguides, these field configurations are called *modes*. Figure 6-7 illustrates several shapes of cavities and the dominant mode of each.

Of the types shown in the figure, the cylindrical type of cavity (Fig. 6-7B) is useful in wavemeters and other frequency-measuring devices. The cylindrical ring and doughnut (Fig. 6-7C and 6-7D) are used in many microwave oscillators as the frequency-determining element. The waveguide section (Fig. 6-7F) is used in some radar systems as a mixing chamber for combining signals from two sources. The cube (Fig. 6-7A) is a special type of waveguide section. The sphere (Fig. 6-7C), although not in common use, can be used for mixing multiple inputs, for changing the mode of operation, for impedance matching, or for extracting multiple outputs.

Note that in each type of cavity shown, the voltage is represented by E-lines between the top and bottom of the cavity. The current, because of skin effect, flows in a thin layer on the surface of the cavity. The H-field is strong where the current is high; the strongest H-field is at the vertical walls of the cavity, and it diminishes toward the center, where the current is zero. This characteristic results from the presence of standing waves, similar to the action on a quarter-wave section of transmission line or waveguide. The *E*-field is maximum at the center and decreases to zero at the edge where the vertical wall is a short circuit to the voltage.

The modes in a cavity are identified by the same numbering system that is used with waveguides, except that a third subscript is used to indicate the number of patterns of the transverse field along the axis of the cavity (perpendicular to the transverse field). For example, the cylindrical cavity shown in Figure 6-7B is a form of circular waveguide. The axis is the center of the circle. The transverse field is the magnetic field; therefore, it is *TM* (transverse

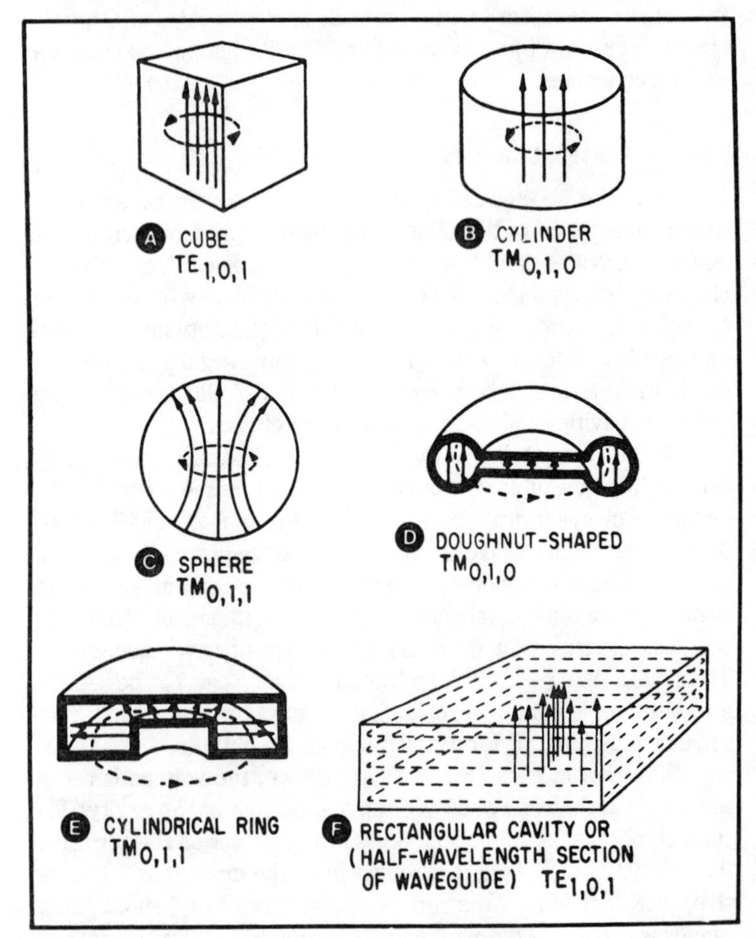

Fig. 6-7. Fields in resonant cavities.

magnetic). Around the circumference there is a constant magnetic field. (The *H*-lines are parallel to the circumference.) Therefore, the first subscript is 0. The distance across the diameter is one-half wave; thus the second subscript is 1. Through the center, along the axis, the *H*-field strength is a constant zero; this makes the third subscript 0. The complete description of the mode is $TM_{0,1,0}$.

When a section of waveguide is closed on both ends in the form of a rectangular cavity, as shown in Fig. 6-7F, standing waves are set up and resonance occurs. The simple mode in this cavity is the same as the dominant mode of a rectangular waveguide; that is, it is TE_{10}. The third subscript of the mode, which is determined by a plane perpendicular to the *E*-field, is 1. Thus the complete description of the simple mode in the rectangular cavity is $TE_{1,0,1}$ (transverse electric).

Exciting the Cavity

A cavity resonator can be excited in any of three predominant methods. These methods are illustrated in Fig. 6-8.

The first method, Fig. 6-8A, is by inserting a probe into the cavity. The voltage applied to the probe sets up an *E*-field with the *E*-lines parallel to the probe. As the *E*-field expands, it in turn sets up an *H*-field. These fields expand and collapse; the wavefronts travel through the cavity and are reflected from the end. If the physical dimensions are correct, these wavefronts will return at the proper time to reinforce the original fields and produce standing waves. Due to the high *Q* of the cavity, these oscillations will sustain themselves for an appreciable time, thus requiring only occasional excitation.

Another method, Fig. 6-8B, uses a magnetic loop. The loop is placed in a region where the magnetic field will be located. The currents in the loop set up an *H*-field within the cavity. The expanding and collapsing of the wavefronts combine to maintain each other if the size of the cavity is correct.

Either of the above methods also can be used to extract energy from the cavity.

A third method, Fig. 6-8C, is used with a cylindrical ring-type cavity. The energy is placed into the cavity by clouds of electrons, which are virtually shot through the holes in the center of a perforated plate. As each cloud passes through, it creates a disturbance in the space inside the cavity until a field is set up. It can be said that the cloud of electrons produces a voltage difference between the perforated plates and thus sets up an *E*-field.

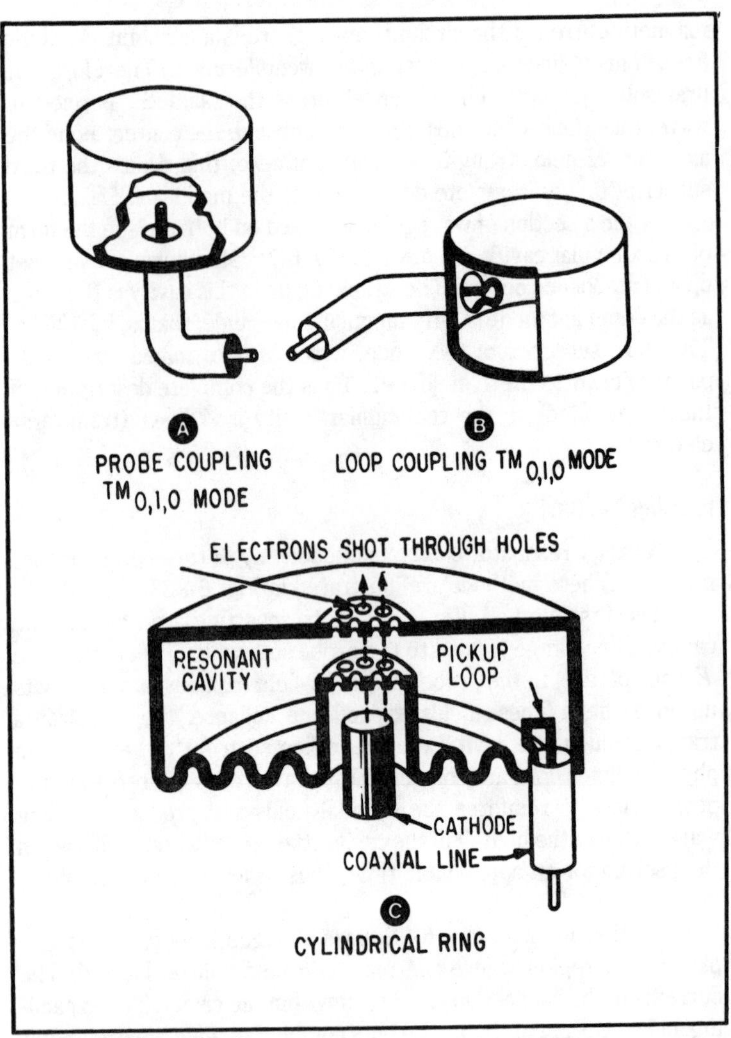

Fig. 6-8. Exciting the cavity.

Tuning the Cavity

Three methods for adjusting the resonant frequency in a cavity are shown in Fig. 6-9. One method uses a cylindrical cavity with an adjustable disk. When a $TE_{0,1,1}$ mode is used, the size of the cylinder can be changed along the axis to change the resonant frequency. The smaller the volume of the cavity, the higher the resonant frequency. The movement of the disk can be calibrated in terms of frequency.

Usually, in the case of high-frequency equipment, a micrometer adjustment is used to change the position of the disk.

A second method of tuning a cavity employs threaded plugs which are inserted in the side of the cavity. The plug reduces the length of the magnetic field in the cavity in a manner similar to reducing the inductance of the tuned circuit. The deeper the plug extends into the cavity, the higher the frequency.

In the third method, the interior of the cavity, which is a part of the interior of a vacuum tube, is sealed and evacuated. In this method, a frequency change occurs whenever the top and bottom of the cavity are moved away from (or toward) each other. This is accomplished by turning a screw that passes through two bowed struts. As the distance varies, the volume and the capacitance between the top and bottom of the cavity are changed. As the change in

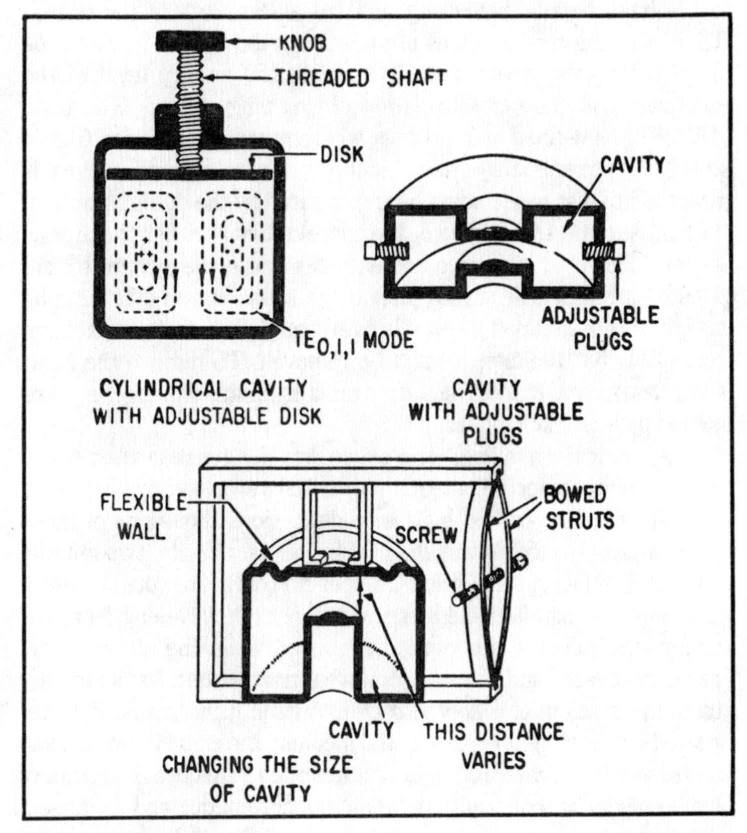

Fig. 6-9. Tuning the cavity.

capacitance is the chief result of the change in the distance between the plates, the resonant frequency is proportional to the distance from the top to the bottom.

Another way of changing the frequency (not shown) is by changing the method of exciting the circuit. This can be done by making the exciting element either capacitive or inductive. This can occur either accidentally because of improperly tuned circuits, or deliberately as a means of tuning the cavity.

CAVITIES AT OPTICAL WAVELENGTHS

A knowledge of microwave cavities as determining the frequency of oscillations, phase relationships, etc., is useful in understanding lasers because lasers also use resonating structures. Most lasers have parallel reflecting surfaces at the ends of the cavity. There also must be a means of preventing the stream of electrons from hitting the cavity wall. To change the energy level of the electron, a method of stimulation or pumping is used. The term *pumping* is defined as a process whereby matter is raised from a lower to a higher energy state. Another major factor in laser design is a method to dissipate excess heat generated by the pumping action. The higher the temperature, the more activity within the atomic system. However, too much activity can cause the system to fail. This is especially true of ruby lasers. Tests have shown that the ruby laser does not function when it is overheated. The cavity must have some heat but the excess must be removed. To remove the heat, some lasers are bathed in a cryogenic medium, while others are cooled by heat exchangers.

At optical wavelengths, a single wavelength resonator would have dimensions inconveniently small. To overcome this problem, a special resonator can be built with dimensions thousands of times greater than the single wavelength. In the laser cavity, two mirrors are installed facing each other. The position of the mirrors is critical; they must be parallel, and a specific number of wavelengths apart. One of the mirrors is completely silvered while the other is only partially silvered and remains somewhat transparent. As the energy from the medium is reflected back into itself it increases. With repeated pulses of light applied to the medium, the energy continues to increase until breakthrough is accomplished. This breakthrough of light energy is coherent, nearly monochromatic, and polarized. These terms coherent, monochromatic, and polarized mean that

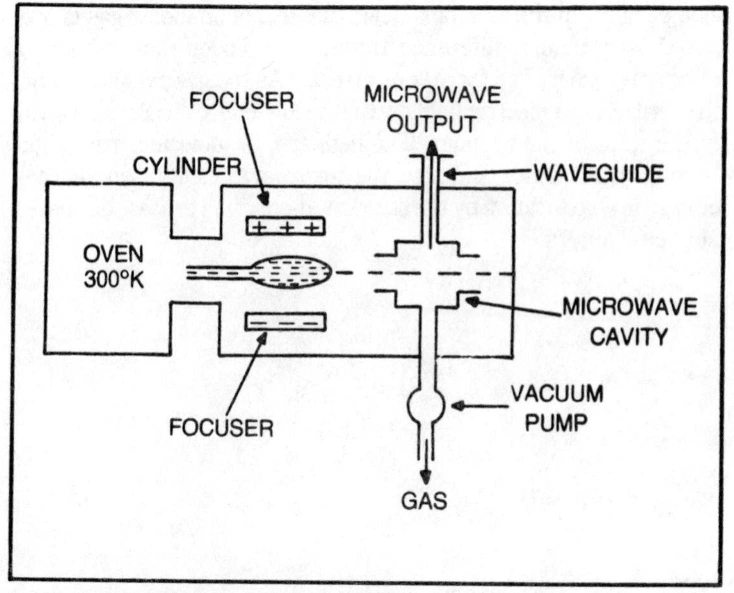

Fig. 6-10. Simplified sketch of the first maser.

there is a fixed phase relationship between the portion of the wave emitted at one instant and the wave emitted after a fixed time interval.

Obtaining population inversion involves the addition of energy to a gas. The process of populating an upper energy level at the expense of a lower one is called ***pumping***. One possible method of pumping between two levels whose difference in energy is E is by supplying electromagnetic energy of the frequency satisfying the relation $hf = E$, thus raising the energy by absorption. This method is efficient only at the start, when the population of the lower level exceeds that of the upper. As equal population is obtained, the number of upward transitions becomes equal to the number of downward transitions, regardless of the pumping energy. To obtain inversion, a more sophisticated technique is necessary, involving at least one intermediate energy level in which the pumped atoms can be stored.

FORERUNNER OF THE MODERN LASER

The ammonia gas maser shown in the simplified sketch in Fig. 6-10 was the first maser and the forerunner of all stimulated-emission devices. In the ammonia beam maser, hot ammonia gas is propelled by pressure difference through a cylinder that contains a microwave cavity and focuser electrodes. As the gas passes through the focuser, the electric field draws the low-energy molecules away to the sides of the cylinder. The high-energy molecules pass right through the focuser and into the microwave cavity, where they engage in amplification by the stimulated-emission process discussed in later chapters.

The Ruby Laser

Now that you have some theoretical background, you can finally start looking at some actual lasers and how they work.

REQUIREMENTS OF A LASER

There are many different types of lasers, but they must all meet the three following requirements:

1. An optically transparent fluorescent material must be used. This material must be pure and free of anything that might interfere with the passage of the stimulated emission.
2. The majority of electrons in the transmission material must be in an excited state, rather than the ground state.
3. Some of the radiation from the stimulated emission must be kept within the transmission material to aid in a chain reaction of photon emission. This is a form of feedback, similar in concept to the electronic signal feedback used in an oscillator circuit.

To understand just what these requirements mean, and to see how they can be met, we will examine the ruby laser in some detail.

USE OF RUBY IN LASERS

As you should recall from Chapter 1, the world's first practical laser was a ruby laser. In hindsight, this isn't at all surprising, since the ruby laser is one of the simplest types.

In its simplest form, the ruby laser is shown in Fig. 7-1. Figure 7-1A is the conventional drawing, while Fig. 7-1B is more familiar to you as a schematic representation. The ruby (aluminum oxide) plus a few chromium atoms (0.05%) sparsely located throughout the aluminum oxide is the material most commonly used. The heart of the device is the cylindrical ruby crystal (*a*), around which is the helical flash tube (*b*). Mirrors at each end (*c* and *d*) reflect the light back and forth through the crystal, and the laser beam emerges through one of the mirrors which is only partly silvered.

Ruby is a form of sapphire (aluminum oxide), containing a trace of chromium. Ruby, like a number of other materials, is naturally fluorescent. That is, it glows when illuminated by light. This fulfills the first requirement for laser action.

When a chromium atom is exposed to a light source containing blue or green wavelengths, one of the outermost electrons will instantly jump to a higher energy level, or an excited state. In this case, there are two possible excited states the electron can jump to.

After a brief period (about 100 nanoseconds), the excited elec-

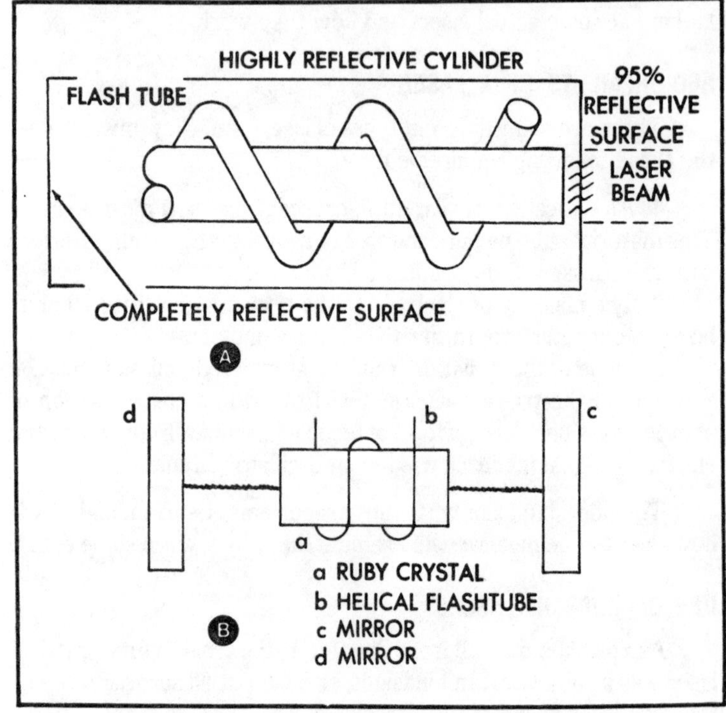

Fig. 7-1. Simple ruby laser.

tron falls back to an intermediate energy level which is lower than the fully excited state but higher than the nominal (unexcited) ground state. This intermediate state is held for a period lasting up to several milliseconds. This period is known as the *fluorescent lifetime*. While in this condition the chromium atom is said to be in a *metastable state*.

After the fluorescent lifetime period is over, the excited electron spontaneously falls back to the original unexcited (ground) state.

When it jumped up to the higher energy level, the electron absorbed energy. This excess energy has to go somewhere when the electron drops back to the ground state. In nonfluorescent substances, this excess energy is given off almost entirely as heat. But in ruby and other fluorescent materials, a portion (often the larger portion) of this absorbed energy is emitted as a photon of light.

In the case of ruby, the emitted photons have a wavelength of 694.3 nanometers. This gives the emitted light a deep red color.

If a sufficient number of electrons is excited, a chain reaction takes place. When electron A drops back to its ground state, it emits a photon, which strikes electron B. If electron B is still in an excited state when it is struck by photon A, it will immediately fall back into its ground state, emitting a second photon that is in perfect phase with the first. If this sequence happens enough, an amplification process called *stimulated emission* occurs. This is at the heart of laser action. Remember that the word laser is actually an acronym standing for "Light Amplification by the Stimulated Emission of Radiation."

Under ordinary conditions, stimulated emission does not occur. If an emitted photon strikes an unexcited electron, the result is just ordinary fluorescence.

Stimulated emission can be assured by placing the majority of electrons in an excited state. Then the odds are very good that an emitted electron will strike an electron while it is in an excited state. Of course, this is the second requirement for laser action described earlier. This requirement can be met by briefly bathing the ruby in a very intense light. In the ruby laser of Fig. 7-1, the helical flash tube is a device for producing this intense light. The ruby absorbs energy from the flash tube, and in a very short time (thousandths of a second) emits it, some energy in the form of light and the rest as heat.

A small part of the light energy produced by the ruby consists of the red beams traveling parallel to its axis. This energy is reflected back and forth by the mirrors so that it passes through the crystal many times. As it passes, it is amplified; that is, it picks up more energy from the ruby. This energy travels in the form of red light along the beam as the intensity continues to build.

Since the ends of the ruby rod are mirrored, most of the emitted photons (light) bounce back and forth within the rod, causing more and more stimulated emission. Some of the emitted light escapes through a small opening in one of the end mirrors. This escaping light is focused into a very tight, very intense beam.

This entrapment of most of the emitted photons by the end mirrors fulfills the third and final requirement of laser action.

THE RESONANT CAVITY

Since the essential property of a gas is that the constituents do not interact with one another, a lightly doped (much less than 1%) crystal is essentially a gas of dopant atoms in a rather special container. Their energy levels are modified by the presence of the host material. A simplified energy level diagram of the chromium ion in a ruby crystal is shown in Fig. 7-2. The intermediate level (3) is metastable with a lifetime of approximately 10^{-3} second.

By examination of Fig. 7-2, you can see that a pumping light of frequency f_{12} causes a transition between levels 1 and 2. The atoms in the excited (2) state can return to the ground state spontaneously, either directly or by first stopping at the metastable state. Because the lifetime of the metastable state is 100,000 times longer than that of state 2, the atoms which fall there can be considered almost stationary. The rate at which atoms find themselves in state 3 is proportional to the rate at which they arrive in state 2, which, in turn, is proportional to the pumping power and independent of the population. If sufficient pumping power is supplied, the population of state 3 grows at the expense of state 1 and population inversion is obtained.

As long as the population is inverted, the ruby can be an amplifier for radiation of frequency f_{31}, and, as with any amplifier, adding a positive feedback loop can cause sustained oscillation. In this case, the positive feedback is the return of some of the output light (f_{31} radiation) into the ruby. This can easily be accomplished with mirrors. By making use of the geometry of the mirrors so the feedback is directional, a resonant cavity is formed. The amplifier radiation, referred to as photon amplification, builds up in a standing-wave pattern that is familiar to you from your study of microwaves (Chapter 6).

The resonant cavity formed in the crystal itself is made possible by carefully grinding and polishing the ruby, then silvering its ends.

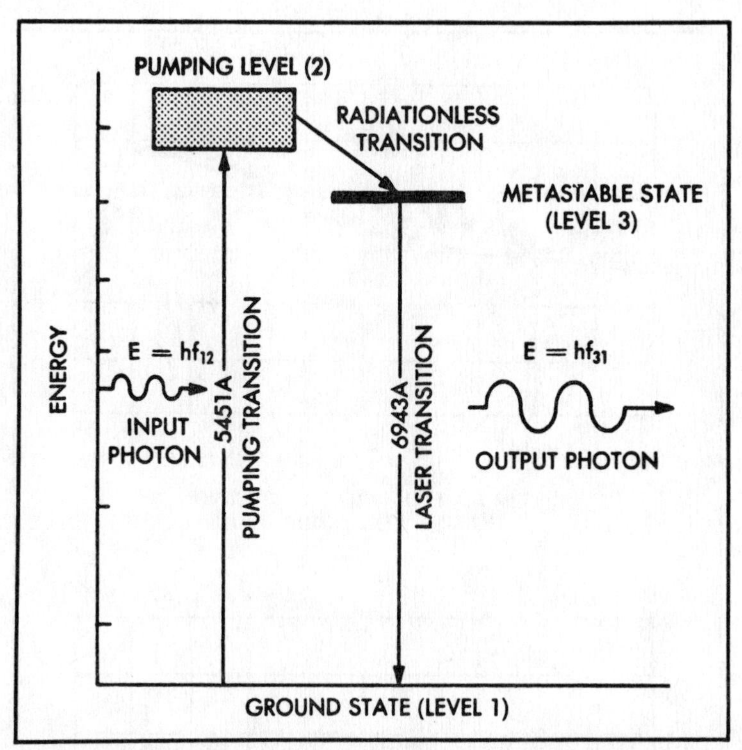

Fig. 7-2. Three levels of chromium ion in ruby rod (associated with pumping and laser action).

Because of the shortness of optical wavelengths, an essential difference exists between the crystal cavity and the more familiar microwave cavity. Calculating the wavelength from the energy-level diagram, the wavelength corresponding to f_{31} radiation is found to be 6943 angstroms in a vacuum. A ruby ground to form a cavity 7.3 centimeters long (a typical size) has 100,000 nodes in the standing wave, and is resonant for every frequency that satisfies the standing wave condition $f_o = n\lambda/2$, where λ is the wavelength and n is an integer. For example, taking n as 10^5, the difference between resonant wavelengths, $\Delta\lambda$, is given by $(\Delta\lambda/\Delta) = (\Delta f/f) = 1/n = 10^{-5}$. The cavity is resonant for a large number of frequencies immediately around f_{31}, instead of being resonant for only one particular frequency, as in the microwave case.

The pumping power required to obtain population inversion for a reasonably sized crystal is considerable. This population inversion can be accomplished only in brief bursts of light from a flash lamp. First, the operating time of the ruby laser is limited to a couple of

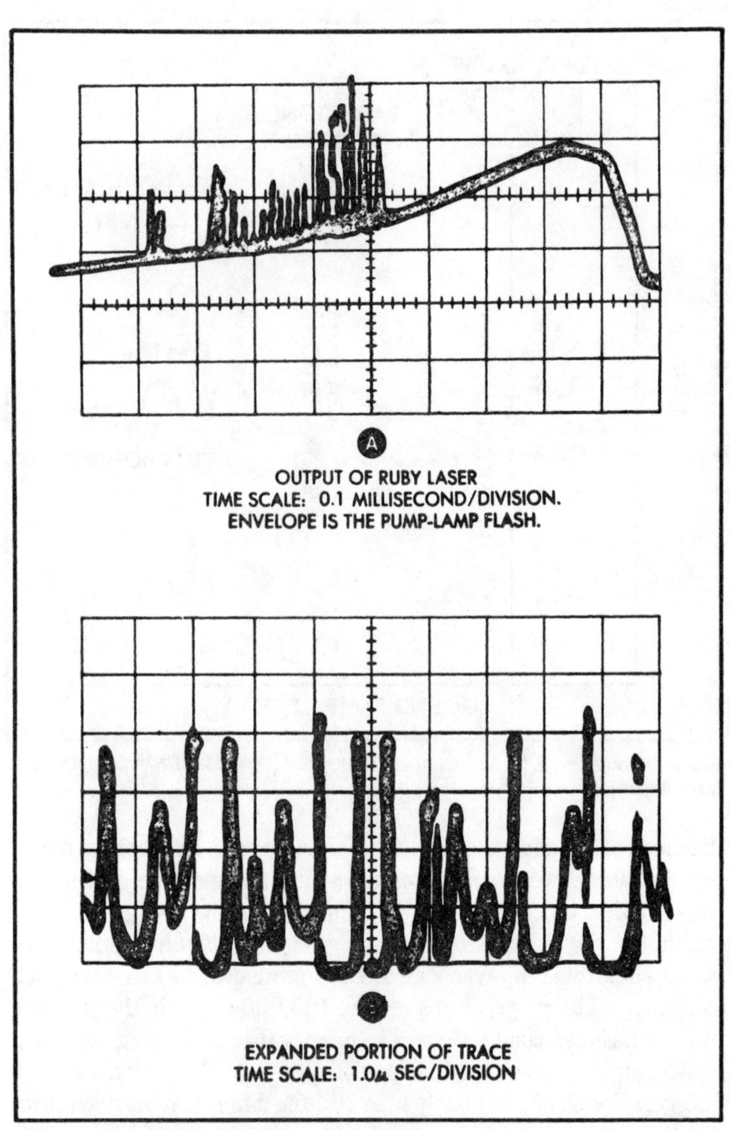

Fig. 7-3. Output of a ruby laser.

milliseconds. Second, while the ruby is operating as a laser, the metastable stage is being depopulated by stimulated emission, and very quickly (in 10^{-6} second) outruns the pump. This causes laser action to stop until the pump can again create a population inversion. The output of a ruby laser is composed of a series of irregularly spaced spikes about 10^{-6} second in duration, in an envelope defined by the pump lamp duration. Figures 7-3A and 7-3B show an oscilloscope trace of the output of a photodetector receiving light energy from a ruby laser.

FInally, consider the overall efficiency of the ruby laser by forming a percentage from the ratio of total output of laser light energy to the electrical energy supplied to the pump. This efficiency is less than 1% with most of the lost energy dissipated in heating the ruby crystal. This makes cooling the crystal an important practical consideration.

While the efficiency of a ruby laser is quite low when compared to other types of lasers, it is still a remarkably powerful device. It is no problem for a ruby laser to put out a burst of radiation of 100,000 watts for a brief pulse (say, about 100 microseconds or so). This could be sufficient (with a suitable lens) to punch a hole through a 3/8-inch thick steel plate in one burst—and that's with just 1% efficiency!

8

The Gas Laser

Gas lasers offer more avenues for exploration than solid-state lasers because the atoms are more accessible for stimulation by a variety of means.

To understand the operation of the helium-neon gas laser, a different means of obtaining population inversion, as well as of pumping, must be considered. The modification of the energy level scheme is shown in Fig. 8-1. Note that a fourth or terminal level has been added above the ground state. The population inversion is now obtained between levels 3 and 4. The advantage of the four-level scheme is that the initial population of the terminal level is negligible as compared with the ground state; therefore, inversion is more easily obtained. That is, fewer atoms in state 3 are necessary for its population to exceed that of state 4 than would be required at the ground level. This reduces the pumping power required, and opens up the possibility of pumping by a different method, called electron-collision pumping.

In quantum theory, an analogy is found for an electron beam traversing atoms of gas and that of a photon beam. Only when the kinetic energy of the electrons coincides with the differences in energy between any two levels are atoms induced by collisions with the electrons to make transitions between these levels. As in the electromagnetic case, the most probable direction of the transition depends only on the relative populations of the states. Rather than the energy from a light beam, the energy of accelerated electrons can

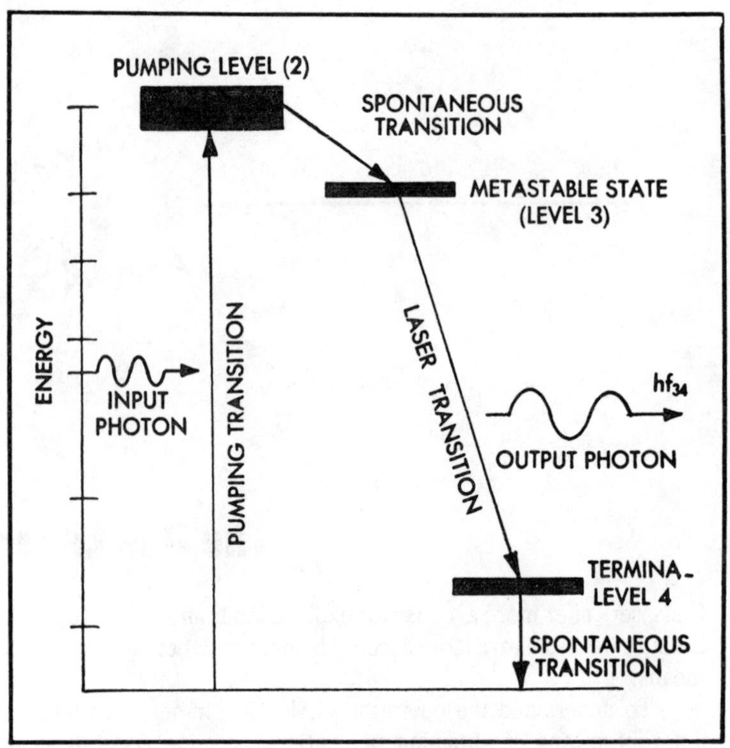

Fig. 8-1. Energy-level diagram of a four-level gas laser.

be used to pump atoms of gas in a discharge tube. These electrons can be the result of a glow discharge when rf energy is applied to the discharge tube. The advantage of this method is that energy can be maintained constant over extended periods of time to obtain a continuous laser output.

The first successful operation of a laser by collision pumping required the presence of two gases, such as helium and neon, in the discharge tube to realize the proper energy scheme. A slight digression is necessary to consider the transfer of energy between atoms of different gases. A sort of resonance phenomenon is encountered where energy transfer proceeds only when an energy gap is shared. That is, an atom of one type of gas (A) in a given energy level ($2A$) can transfer its energy to an atom of another type (B) in a stage ($1B$) via a collision if, and only if, there exists energy levels $1A$ and $2B$ of such value that $E_{2A} - E_{1A} = E_{1B} - E_{2B}$ (see Fig. 8-2). As before, the probability of the transition direction is determined by the population of the levels.

In the gas laser, the higher and lower energy levels are in different gases, and their population can be modified by changing the relative concentration of the different gases in the discharge tube. This additional control was fundamental in achieving the first observed laser action in a gaseous mixture of helium and neon.

Figure 8-3 shows the pertinent sections of the energy-level diagrams for helium and neon with the transitions indicated. Note that when level 3 is well populated by pumping energy, there is amplification for two different frequencies, f_{34} and f_{34}'. Note also that the alternate route 1, 2′, 3′, 4′ should result in amplification at the frequency f_{34}'. Laser action has been observed at all of these frequencies in helium-neon mixtures. Selection of oscillation between these frequencies is accomplished by using feedback mirrors with reflectivities at different frequencies.

The most distinctive difference between the helium-neon laser and the ruby laser is in their outputs. The ruby laser output is typified by irregular spiking, while the helium-neon is capable of a continuous wave of extremely narrow bandwidth. This distinction of emissions has determined the field of application of the particular laser. Equip-

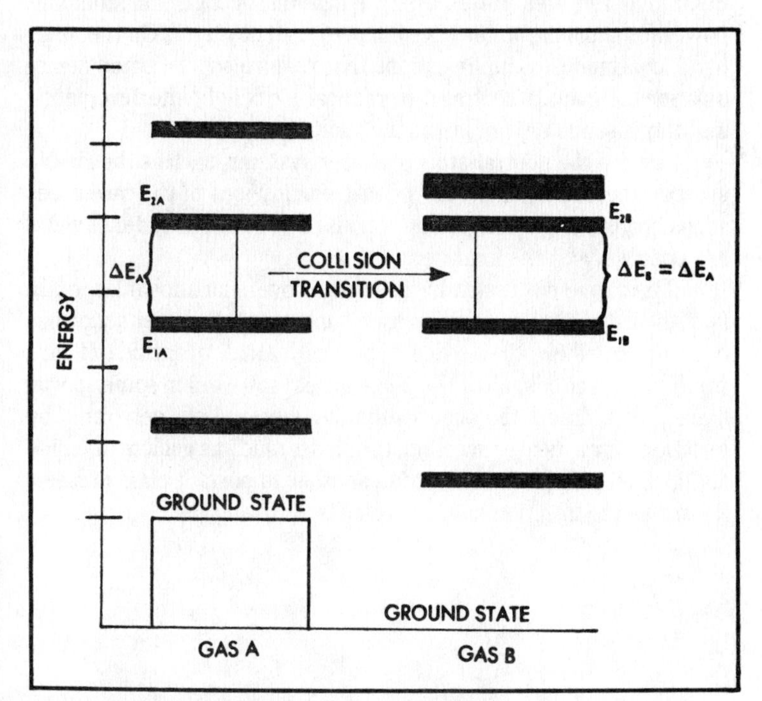

Fig. 8-2. Resonant transfer by collision of gas atoms.

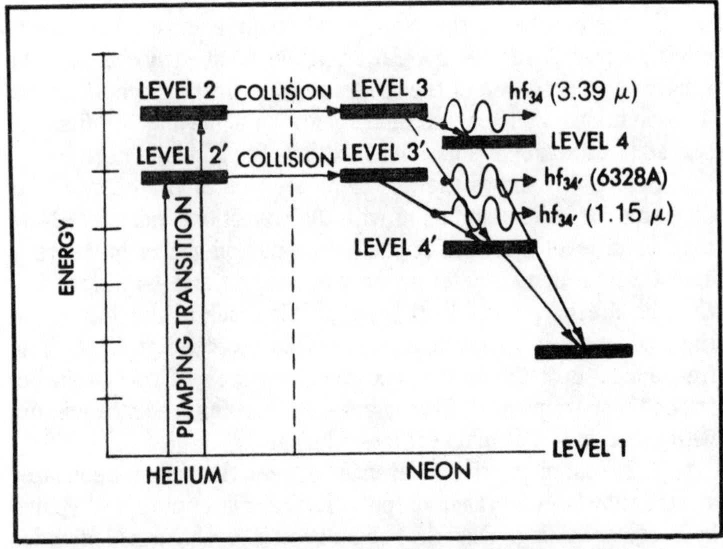

Fig. 8-3. Energy-level diagram of helium and neon during laser action.

ment that requires pulsed-typed emissions of high instantaneous power for short durations have adopted the ruby laser. On the other hand, continuous communications that require sophisticated demodulation techniques, such as the practical use of light interferometry, find the gaseous type of lasers well suited.

Besides the neutral-atom gas laser systems, such as the He-Ne system, there are ion gas lasers that employ ions of the rare gases argon, krypton, and xenon. Continuous powers on the order of watts are possible with the ion lasers.

The above discussion by no means covers all known lasers. In fact, the list of laser materials grows almost weekly. Laser action has been observed in other solids doped with small quantities of rare earth ions, in almost all of the noble gases, and even in some special liquids. In addition, the observation of coherent light generated by injection currents in semiconductor diodes, such as gallium arsenide and gallium phosphide, has added another important class of laser, known as the injection laser, to this fast-growing field.

9

Laser Energy and Emission Characteristics

Laser operation requires an active material that produces emission of radiation, an excitation source that pumps power into the active material, and a resonant structure. Both solid-state and gas lasers have these characteristics. The functioning of the solid-state lasers developed up to now is essentially the same as that of the original ruby laser. Figure 9-1 shows a block diagram of a typical ruby laser. The active material is ruby, the excitation source is a xenon flash tube, and the resonant structure is formed by a ruby rod whose ends are reflecting mirrors. One end of the rod has a heavy silver coat that forms an opaque mirror; the other end has a light silver coat that forms a 92% reflecting mirror.

Although the complete physical and mathematical description of laser action is complicated, it is possible to get a simplified picture of laser action by relating this action to the energy-level diagram for a Cr^{+++} (chromium ion)-doped, pink-ruby crystal (Cr_2O_3: Al_2O_3), the material used in the ruby laser. The heights of black bars 1 and 2 and of the area shown in crosshatching 3 indicate the possible energies that a Cr^{+++} ion can have. Energy is in units of 10^3 wavelengths per centimeter; one wavelength per centimeter ($1\ cm^{-1}$) is equivalent to 1.9858×10^{-16} erg. In their normal condition (the ground state), the Cr^{+++} ions have zero energy. This condition is indicated by level 1.

If light photons having a wavelength of 5600 angstroms irradiate the ruby crystal, they will raise the energies of some Cr^{+++} ions

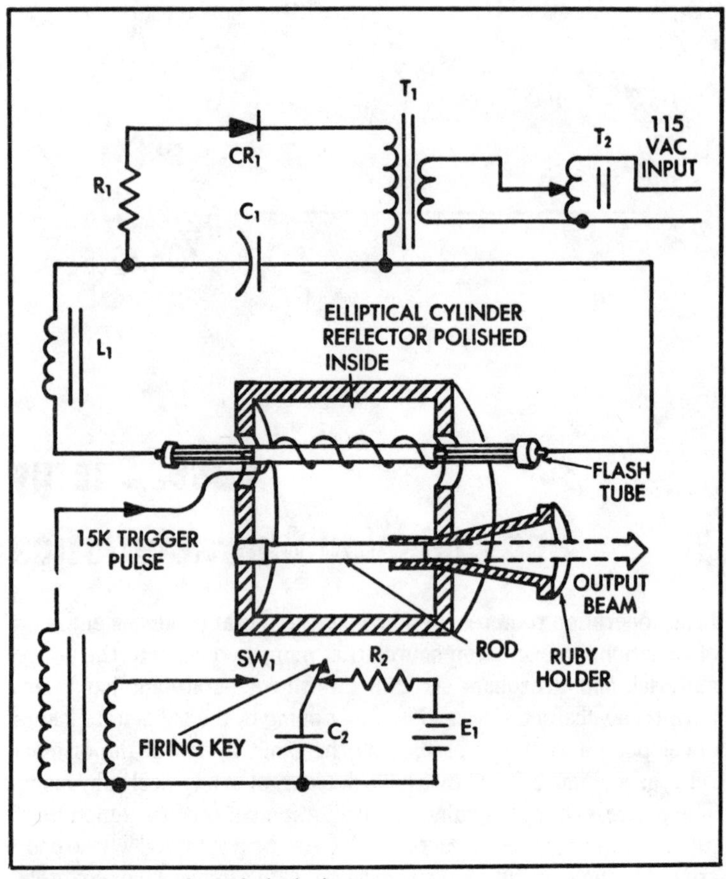

Fig. 9-1. Diagram of a typical ruby laser.

to various energy levels lying in the absorption band of energies indicated by 3. Flash tubes supply this irradiating light, along with light composed of many other wavelengths. The left-hand arrow, W_{13}, going from level 1 to band 3 indicates the increase in energy acquired by a Cr^{+++} ion when it absorbs a 5600-angstrom light photon. The use of light excitation to raise the energy level of atoms to higher levels is called *optical pumping*. After short but finite times elapse (relaxation times), some of the Cr^{+++} ions in band 3 drop back to level 1 (shown by A_{31}), and some drop to level 2 (shown by S_{32}). The rate at which Cr^{+++} ions drop to level 2 is greater than the rate at which they drop to level 1. The Cr^{+++} ions in energy level 2 hold their energy for a short time before they drop to level 1. The rate that ions go from level 2 to level 1 (A_{21}) is less than the rate at which Cr^{+++} ions go from level 1 to level 3. Optical pumping builds up the number of ions having level 2 energies to a greater number than the number of ions having level 1 energies. In other words, the Cr^{+++} populations of levels 1 and 2 are inverted from their normal relationship. This population inversion is essential for producing stimulated emission.

In dropping from level 2 to level 1, Cr^{+++} ions radiate light. Level 2 ($2E$) is actually composed of two levels, levels E and $2A$, which emit radiation lines R_1 and R_2, respectively. If conditions were not completely correct for achieving laser action (for example, if an insufficient amount of excitation were applied), R_1 radiation would be spontaneous radiation rather than stimulated radiation, and would include a much broader band of wavelengths than the stimulated emission (Figs. 9-2B and 9-2C). In both cases, the center frequency of the R_1 and R_2 energy radiated when ions drop from level 2 to level 1 is calculated from $v = (E_2 - E_1)/h$, where v is the frequency; E_2 is the energy at the center of level E (for R_1), or level $2A$ (for R_2); E_1 is a constant representing the first energy level; and h is Planck's constant. The resonating character of laser action enhances radiation at the central wavelength of R_1 and diminishes other radiation. Arrow A_{21} in Fig. 9-2A indicates the spontaneous radiation of R_1 and R_2 that will be emitted if laser operating conditions are not correct. The broken arrow indicates the laser output, which is composed of coherent laser radiation at R_1 and spontaneous (incoherent) radiation at both R_1 and R_2.

The simplified sketches shown in Fig. 9-3 illustrate sequences of laser action. At the instant that pumping light is applied, all Cr^{+++} ions are in the ground state; the unshaded circles indicate this state. Optical pumping raises some Cr^{+++} ions to level E (Fig. 9-3A). The black circles in Fig. 9-3B indicate ions that have been pumped up to level E. Some Cr^{+++} ions drop to level 1, radiating photons that have various wavelengths centered about the central wavelength of R_1. (Consider a photon as a bundle of light energy that has wavelike properties.) Figure 9-3C shows ion A, which is dropping to level 1, spontaneously emitting radiation. In these simplified sketches, ion A is the first ion (and the only one that is shown) to emit R_1 radiation spontaneously; that is, without being stimulated by R_1 radiation. The radiated photon tends to stimulate radiation of the same wavelength from other Cr^{+++} ions of level E that are in its path. This is indicated in Figs. 9-3C and 9-3D. Assume that incident radiation from ion A has the same wavelength of the strongest R_1 emission. Incident

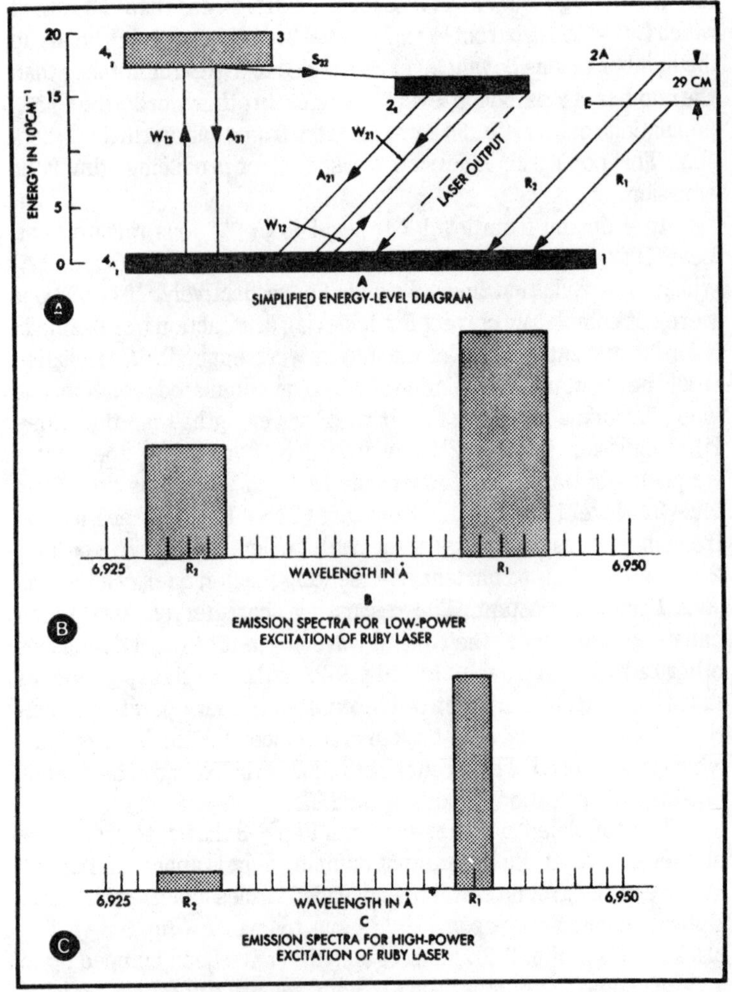

Fig. 9-2. Energy-level diagram for a ruby laser.

photon a is reinforced by stimulated photons, b, c, and d in a precise phase relationship, as indicated by light rays $a + b$, $a + c$, and $a + d$ in Fig. 9-3D. The opaque mirror reflects $a + b$ back into the ruby cavity, but $a + c$ passes through the side wall and is lost. The cavity enhances radiation propagated parallel to the axis of the ruby rod, and minimizes radiation going in other directions. Because of the amplification caused by photons of the same wavelength, rays composed of photons of the center wavelength of the R_1 line, which is the

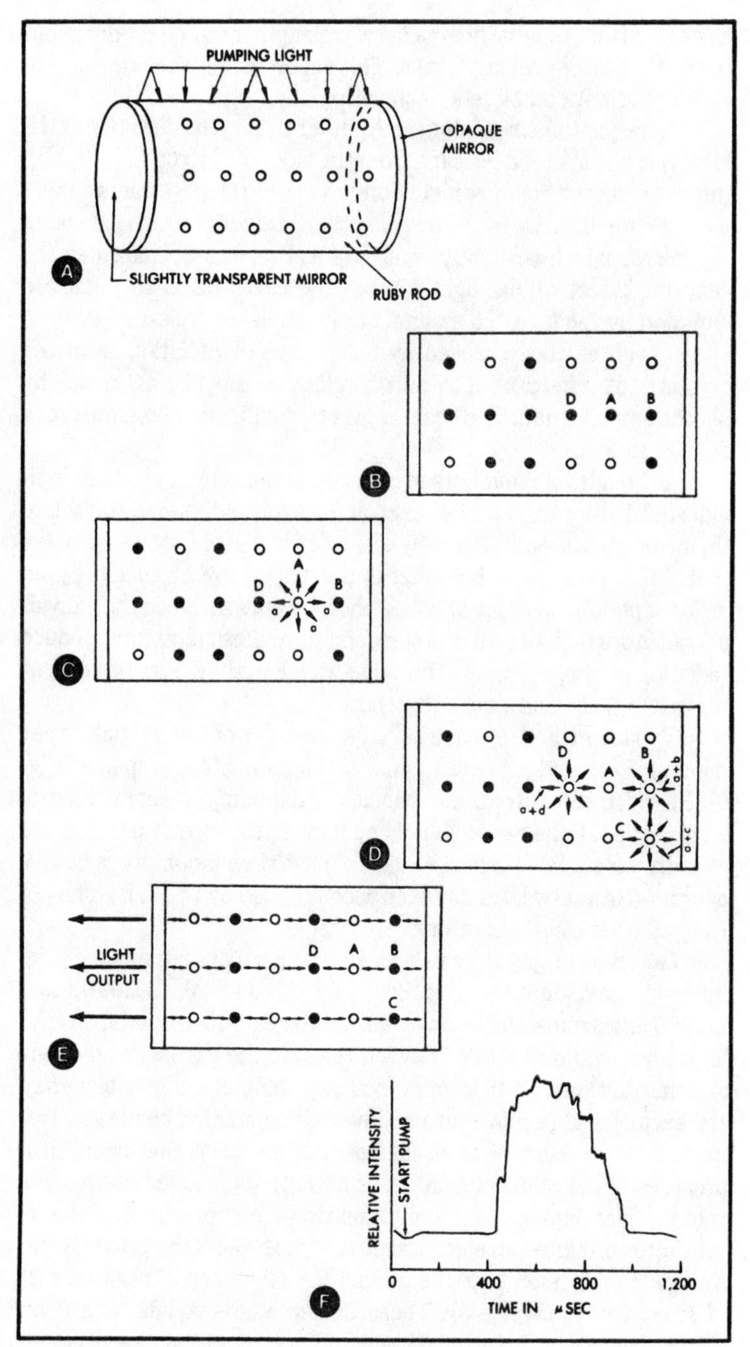

Fig. 9-3. Emitted-radiation sequence of a laser beam.

strongest (that is, most prevalent) wavelength, become predominant over other R_1 wavelength rays. This action makes the laser output highly, but not completely, monochromatic.

Since photons traversing paths other than in the direction of the long axis of the crystal escape from the sides of the rod (Fig. 9-3D), the laser output beam is highly directional. Proton streams reflect back and forth between the end mirrors and emerge from that end of the mirror which is slightly transparent. Figure 9-3E indicates the cohering effect of the light field in the cavity used to stimulate emission by the ions. To maintain the stimulated emission process, losses such as those caused by the escape of radiation from the crystal and losses caused by mirror reflection must be overcome by photon amplification. The beam angle of the ruby laser is on the order of 0.01 radian.

Although vast numbers of ions within the ruby crystal are individual radiators of photons, laser action causes the ions to radiate their energy in step (coherently). The key action of the laser process that produces in-step radiation is the triggering of an ion by an incident photon, in order to emit a photon in-phase. Since the individual radiators radiate nearly in step, and since these radiators produce radiation of approximately the same wavelength, the laser's output beam has space and time coherence.

Typical PRF rates have been on the order of several pulses per minute. Each output pulse contains large-amplitude spikes (Fig. 9-3F) which result from the inability of the pump to supply energy level 2 at a rate fast enough to keep up with the rate at which these ions drop from level 2 during the stimulated emission process. CW operation of a ruby laser has been accomplished by Dr. V. Evtuhov of Hughes Research Laboratory.

One type of gas laser provides continuous operation at several infrared wavelengths—11,180, 11,530, 11,600, 11,990, and 12,070 angstroms, the strongest being the 11,530-angstrom output line. The main difference between the way the gas laser functions and the way solid-state lasers function is the method by which they are excited and pumped into an inverted population condition. Helium, at a pressure of one millimeter of mercury, and neon, at a pressure of 0.1 millimeter are sent into the laser tube, whose ends contain flat, parallel, and semitransparent mirrors. A 30-MHz rf generator produces an electrical discharge through the gas mixture, thus raising the energy of the ground state between atoms to the 2^3S energy level (Fig. 9-4). These helium atoms collide with neon atoms that are in the ground state and energy exchanges between

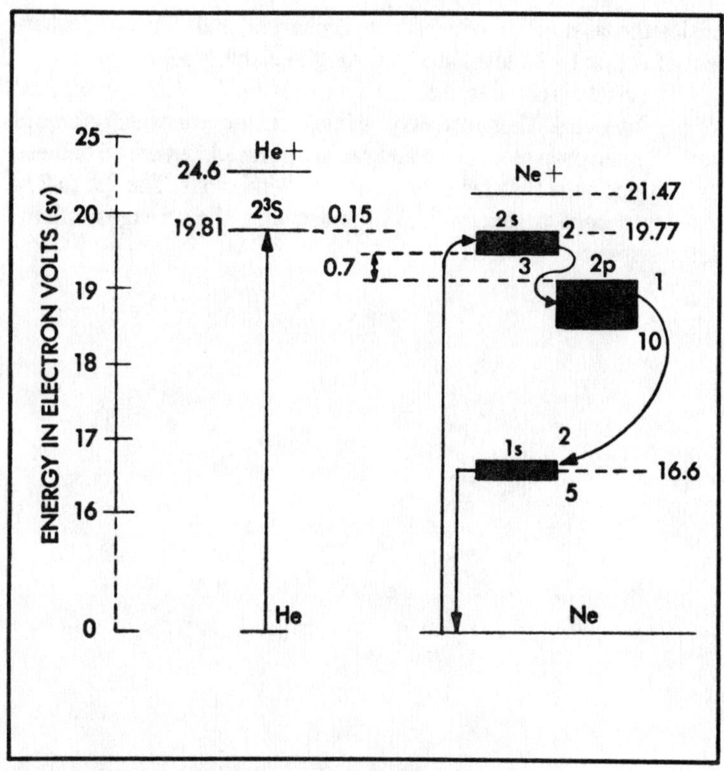

Fig. 9-4. Energy-level diagram of helium and neon gases.

helium and neon atoms take place. Because of the collision, the internal energy of the ground-state neon atom increases to the $2s$ level, which comprises four sublevels, and the internal energy of the 2^3S helium atom drops to zero. This collision process produces $2s$-level neon atoms rather than $2p$- or $1s$-level neon atoms, because the energy-exchange process is such that the least possible amount of change in the total internal energy of the colliding atoms occurs. Since the energy of the $2s$ levels is nearly equal to that of the 2^3S helium level, a 2^3S helium atom readily transfers its internal energy to a $2s$ neon atom, with little of the 2^3S helium atoms's internal energy being converted to kinetic energy. Building up the $2s$ neon population produces a sufficiently large inverted population between levels $2s$ and $2p$ to achieve laser action.

Soon after optical pumping raises the population of the $2s$ level above the population of the $2p$ level, the light field that is built up inside the laser tube cavity becomes coherent, and stimulates coherent emission of 2 *s* level atoms throughout the tube.

Level 2 *s* is composed of four sublevels and level 2 *p* is composed of ten sublevels. There are 30 possible transitions in which atoms in level 2 *s* can drop to level 2 *p* and radiate infrared, but only 5 of these transitions have thus far been stimulated emissions. The $2s_2$ to $2p_4$ energy change produces 11,530 angstroms, the strongest stimulated emission.

10

Laser Classification and Construction

Lasers have been classified largely according to the nature of the material (see Tables 10-1 and 10-2) placed between the two reflecting surfaces. If this material is a solid, such as a rod of ruby crystal, the device is referred to as a solid-state laser. If the material is a gas, such as a mixture of helium and neon enclosed in a quartz tube, the device is termed a gas laser. If the action takes place at the junction of two semiconductor materials, such as a gallium-arsenide (GaAs) diode, the device is termed a semiconductor diode laser. If the material is a liquid or a plastic, such as doped chelates in alcohol, or vinylic resin, the device is designated a liquid or plastic laser, respectively. These latter devices are sometimes referred to as Raman lasers when they are used to study or apply the light-scattering phenomenon known as Raman spectra. These are the extra spectral lines appearing near the prominent lines of the spectrum obtained when a strong light passes through a transparent liquid or gas.

Figure 10-1 illustrates some relative relationships between various laser outputs and the visible electromagnetic spectrum.

A summary of important properties of typical lasers is presented in Table 10-3. The lasers are essentially listed by types in most cases with a variation of laser material.

The Spectra-Physics Model 168 is a fairly typical argon laser. It has an output of up to five watts; a krypton version has an output of up to 0.6 watt. In the multiline mode, the argon laser produces outputs at various discrete wavelengths in the range 457.9 to 514.5 nano-

Table 10-1. Arbitrary Classification of Laser Types.

OPTICALLY PUMPED SOLID DIELECTRIC	Pulsed—low energy; high energy; high repetitive rate—high average power Continuous Crystalline host (sometimes called solid state) Glass host Rare-earth host
GAS LASER	Pulsed Continuous—low power; medium power Arrays Ionized gas—pulsed Continuous—low power; high power R F excitation (pump) DC excitation (pump) Optical excitation (rare)
DIODE LASER	Pulsed—low energy Continuous—low power; high power Array Room temperature Large volume diode laser Electron beam pumped semiconductor laser (not diode)
STORAGE LASER	Giant pulse Q-switched, etc. Passive Q switch (saturation absorption) Storage diode laser
SPECIAL LASERS	Frequency doubled Frequency tripled Raman laser Liquid laser Plastic laser
AMPLIFYING LASER (special cases of basic types)	Oscillator—amplifier for pulsed power Traveling wave laser (nonresonant cavity) Dump amplifier—saturation amplifier Ring laser(s)

Table 10-2. Laser Materials.

HOST	ION	WAVELENGTH (microns)	TEMP. (°K)
INJECTION LASER DIODES			
GaAs		0.8400	77
Ga ($As_{1-x}P_x$)		0.84—0.71	77
InP		0.908	77
InAs		3.1	4-77
SiC		0.5	300
ORGANICS			
Benzophenone—naphthalene		0.47	77
Eu^{3+}—TTA chelate		0.6130	77
GASEOUS LASERS			
He—Ne		0.6328—3.2913	24 or more lines
He		2.0603	
Ne		1.1523—17.88	4 lines
A		1.6180—12.14	5 lines
Kr		1.6900—7.06	9 lines
Xe		2.0261—12.913	2 lines
A—O_2		0.8446	
Ne—O_2		0.8446	
Cs		7.1821	

meter (1 nanometer = 10^{-9} meter). The greatest single-line power outputs are produced at 514.5 and 488.0 nanometer (2 watts and 1.5 watts).

The exciter, shown on the right, provides high-voltage excitation power and control functions. It includes an automatic-start function. The start sequence is initiated 30 seconds after the circuit breaker on the front panel is actuated.

Applications of a CW argon laser of this power include industrial spectroscopy, data recording, biological cell sorting, eye surgery and endoscopic coagulation. An endoscope is a flexible fiber-optic system

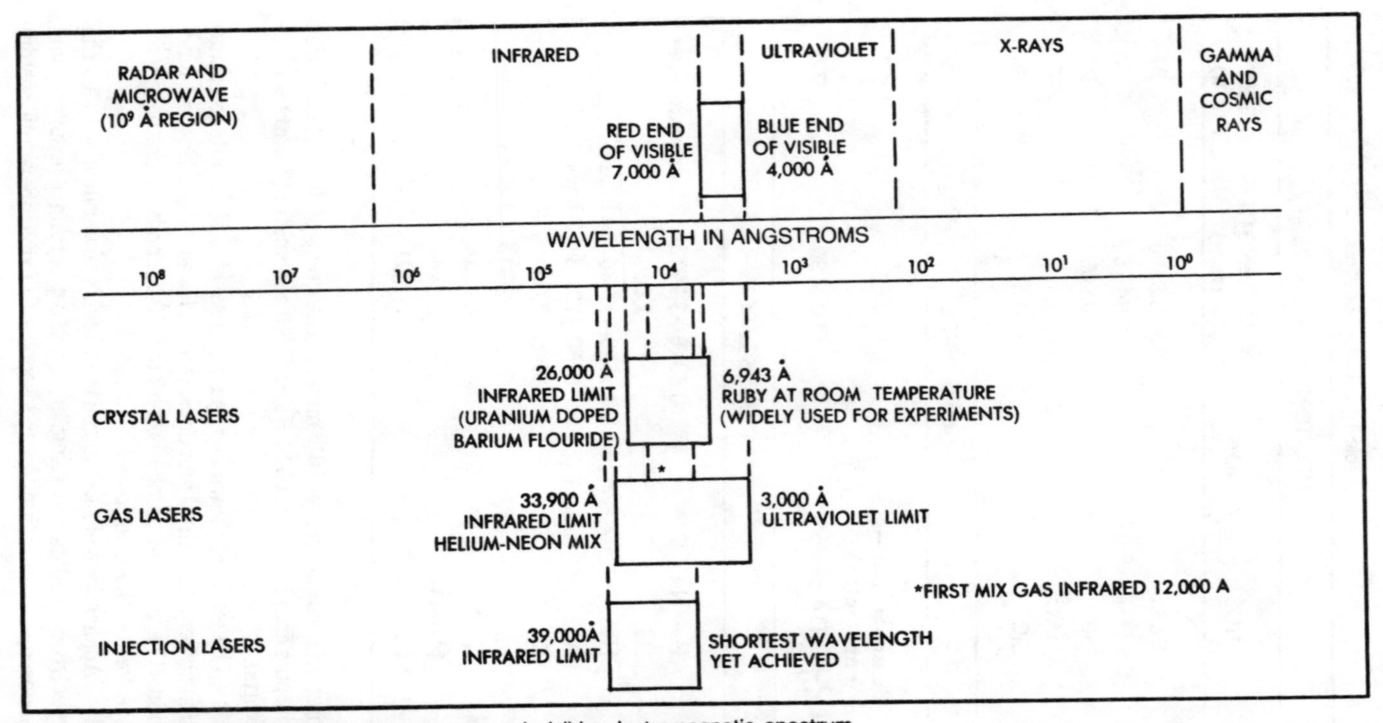

Fig. 10-1. Relationships between laser outputs and visible electromagnetic spectrum.

Table 10-3. Summary of Important Properties of Typical Lasers.

LASER MATERIAL	TYPE	MEANS OF EXCITATION	EMISSION WAVELENGTH	OPERATION MODE	EFFICIENCY OR GAIN	TYPICAL OUTPUT POWER
Ruby	Solid	Optical pumping	6943Å	Pulsed	0.2%	1 to 10 mW
Helium-neon	Gas	Electrical discharge	6328Å 1.15μ 3.39μ	CW	0.05 0.1%/m 5	10 to 100 mW
Helium-xenon	Gas	Electrical discharge	3.5μ 2.026μ	CW	1%/m 15%/m	100 mW
Xenon	Gas	Electrical discharge	3.5μ	CW	5%/m	10 mW
Cesium-helium	Gas	Optical pumping	7.18μ	CW	1%/m	1 to 10 mW
Argon		Electrical discharge	4880Å	CW	15%/m	1 to 10 W
Neodynium in glass	Solid	Optical pumping	1.06μ	CW	.05%	1W
Gallium arsenide	Diode	Injection current	8450Å	CW, Pulsed	10 to 30%	.1 to 1W

that permits visual observation of the interiors of human organs. When combined with a laser, an endoscope can treat gastrointestinal bleeding as well as help the physician to diagnose it. The argon laser emits blue-green light, which is strongly absorbed by blood and bleeding tissue and provides maximum coagulation with minimum damage to surrounding healthy tissue.

The resonating structures (cavities) of most lasers have parallel reflecting surfaces at the ends of the cavities, as shown in Fig. 10-2. The confocal resonator shown in Fig. 10-2B is formed by two spherical reflectors separated by their equal radii of curvature. A confocal resonator has certain advantages over a parallel-plane cavity. It has lower diffraction losses and requires less pumping power. Optical alignment of parallel reflectors is critical, whereas optical alignment of spherical reflectors is not. Figure 10-2C shows a cavity geometry that is designed to attain total internal reflection.

Silver mirrors and mirrors with dielectric coatings have been used as end reflectors in both solid-state and gas lasers. A dielectric coated mirror contains a number of dielectric layers, having different optical matching characteristics. Silver has a lower reflectance than a dielectric coating, and causes relatively higher losses, particularly when used in ruby lasers which have high peak powers. Silver coatings deteriorate with time and use. After several hundred output pulses, the exact number depending on operating conditions, silver mirrors have to be replaced because their deterioration begins to reduce the output significantly. Dielectric coatings do not suffer such deterioration.

Heat affects solid-state lasers more than gas lasers. Tests and observations of extended periods of operation have shown that a ruby laser does not function when it is overheated. Internal heating of the laser material is one of the factors that until recently has prevented CW laser operation of solid-state materials, in spite of the use of such external coolants as liquid nitrogen. Since heat transfer requires time, heat can be trapped within the laser crystal for a long enough period to block continuous laser action. Heating tends to broaden the widths of radiated spectral lines.

The helium-neon gas laser shown in Fig. 10-3 operates at lower power levels than the solid-state lasers developed up to now. Typically, rf excitation power is 50 watts, although it can be as low as 10 watts, and has been reported as high as 80 watts. If too much rf power is applied, laser action will not occur. The minimum discharge-tube length that is necessary for laser action is about 20 centimeters. The rf exciter of the laser runs at 28 MHz (30 MHz

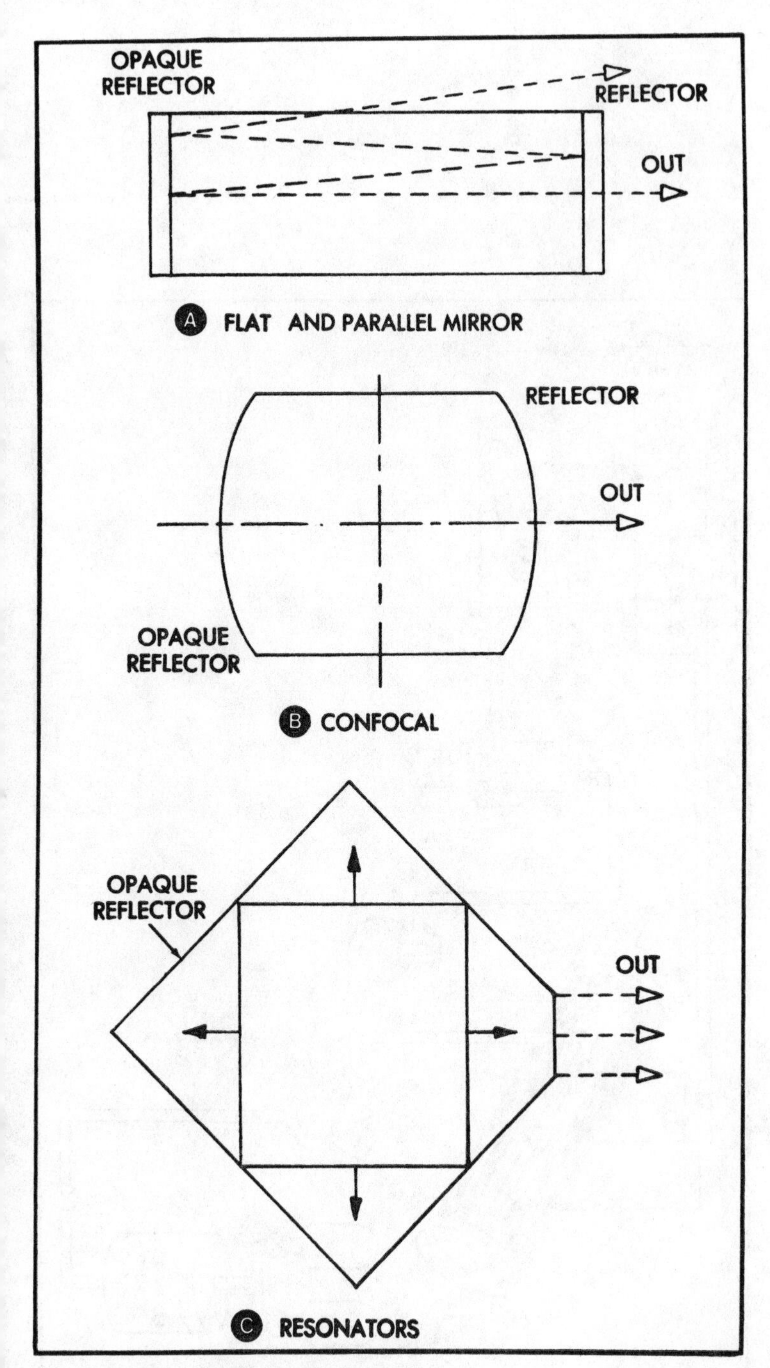

Fig. 10-2. Laser resonant-cavity structures.

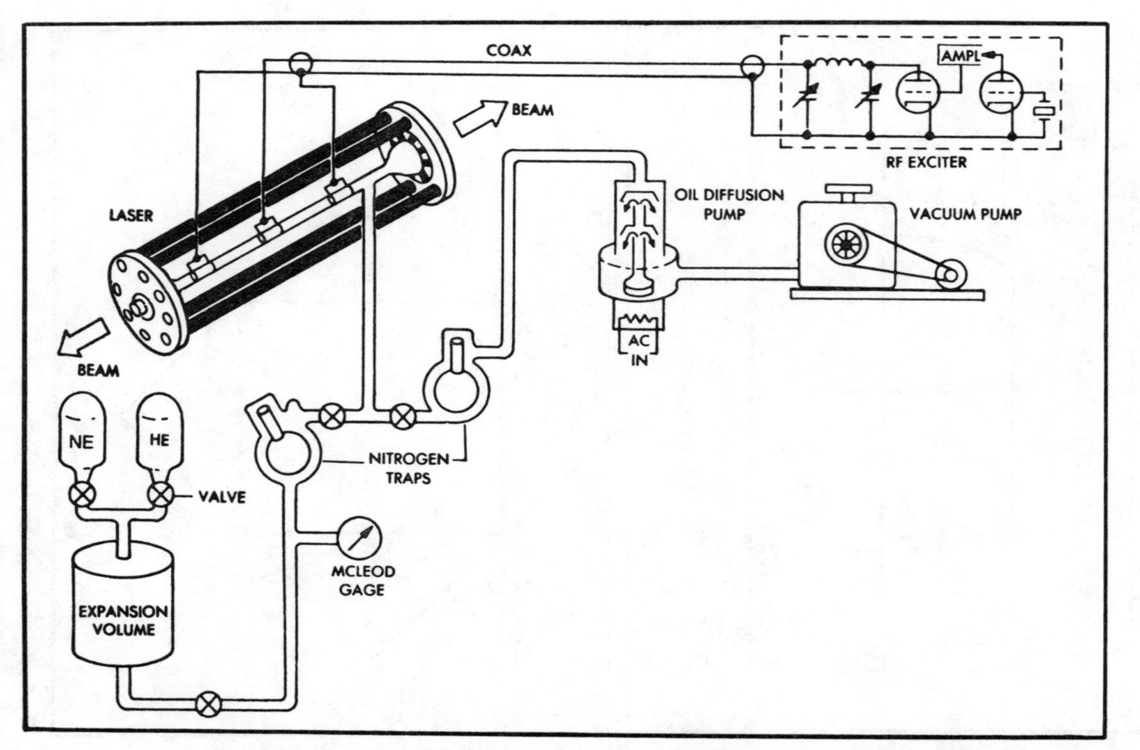

Fig. 10-3. A typical gas laser diagram.

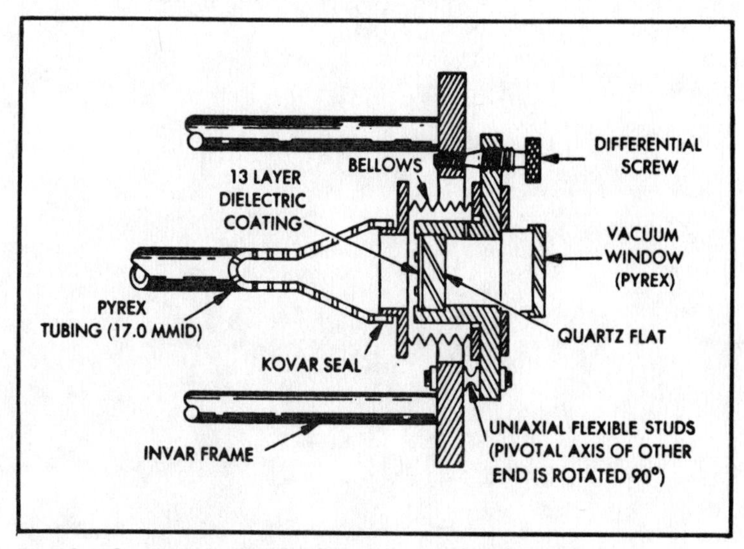

Fig. 10-4. Cross-section details of the end assembly of a gas laser.

excitation has also been reported). The tube is filled with a mixture of helium at 1.0 millimeter of pressure and of neon at 0.2 millimeter. The Pyrex discharge tube (quartz tubes have also been used) is mounted on a low-expansion Invar rod frame. Each of the two end assemblies shown in Fig. 10-4 contains a flexible metal bellows. A differential screw at each end brings the dielectric-coating mirrors into parallel alignment, the differential screw movement providing a resolution of one second of arc. Mirror faces are spaced 1.0 meter apart. Uniaxial flexible studs allow slight positioning adjustments of the end plates. Output windows are sealed with copper gaskets. The window at the end that is not shown in Fig. 10-4 is a quartz flat. A spring-loading mechanism (not shown) provides a vernier adjustment of parallelism. Copper-brazed stainless steel is used throughout the construction.

11

Introduction to Semiconductor Lasers

A semiconductor laser is one in which the laser action takes place at the junction of two semiconductor materials. The gallium-arsenide laser, discussed in the next chapter, is a prime example. This laser is of special interest because it is the type available to the home experimenter. One feature of the gallium-arsenide injection laser is the simple method by which it is pumped to produce laser action. Injection lasers use electrical energy directly, pumping electrons to high energy states by injecting electrons and holes across a pn junction. This chapter introduces the semiconductor pn junction and explains the action of the electrons and holes at the junction.

Another feature of the gallium-arsenide laser is that it operates in the infrared (IR) portion of the electromagnetic spectrum, at a frequency just below that of visible light. The IR portion of the spectrum is divided into three parts: near infrared (NIR), intermediate infrared (IIR), and far infrared (FIR). Gallium-arsenide lasers operate in the NIR part of the IR region. Certain other lasers operate in the IIR and FIR ranges. In the FIR region beyond 15 microns, IR radiation begins to act more like radio emissions than light. There, ordinary optics, even those using such materials as rock salt for prism and lenses, tend to lose their effectiveness, and the physics and physical arrangements of the maser tend to dominate. So it is here, in the IR region, that the borderline between lasers and masers exists. This chapter considers IR radiation and the optical systems and detectors for it.

Injection lasers are very attractive for communications purposes, especially space communications. The lasers themselves are light and compact, and their emissions are sharply focused, resistant to atmospheric absorption, and capable of very fast modulation (which has been achieved at the X-band frequencies—5.2 to 10.9 MHz—used in radar).

SOLID-STATE PHYSICS

The number of electrons in the outermost shell determines the *valence* of the atom. For this reason, the outer of an atom is called the valance shell; and the electrons contained in this shell are called the valence electrons. The valence of an atom determines its ability to gain or lose an electron, which, in turn, determines the chemical and electrical properties of the atom. An atom that is lacking only one or two electrons from its outer shell will easily gain electrons to complete its shell, but a large amount of energy is required to free any of its electrons. An atom having a relatively small number of electrons in its outer shell in comparison to the number of electrons required to fill the shell will easily lose the valence electrons.

The valence shell always refers to the outermost shell, whether it be a major shell or subshell. The copper and sodium atoms each have one electron in the outermost shell. Even though the atomic weights and atomic numbers of copper and sodium are quite different, the atoms are similar in that they both contain one valence electron. Since copper (Cu) and sodium (Na) have one valence electron (refer to Fig. 11-1), they both appear in group one of the periodic table (see Appendix A). Group one designates all elements having one valence electron.

Ions and Ionization

It was mentioned previously that ions do exist and that they are atoms that have assumed a charge. It was stated that there are positive and negative ions. The process whereby an atom acquires a charge will now be discussed.

It is possible to drive one or more electrons out of any of the shells surrounding the nucleus. In the case of incomplete shells, it is also possible to cause one or more additional electrons to become attached to the atom. In either case, whether the atom loses electrons or gains electrons, it is said to be *ionized*. For ionization to take place, there must be a transfer of energy which results in a change in the internal energy of the atom. An atom having more than

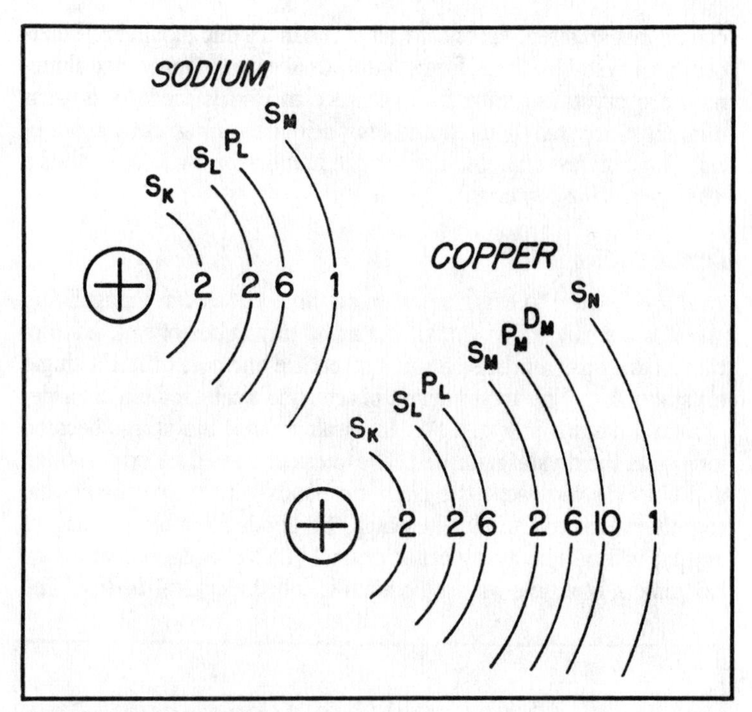

Fig. 11-1. Copper and sodium atoms.

its normal count of electrons is called a *negative ion*. The atom that gives up some of its normal electrons is left with fewer negative charges than positive charges and is called a *positive ion*. Thus, ionization is the process by which an atom loses or gains electrons.

To drive electrons out of the shells of an atom requires that the internal energy of the atom be raised. The amount of energy required to free electrons from an individual atom is called the ionization potential.

The ionization potential necessary to free an electron from an inner shell is much greater than that required to free an electron from a outer shell. Furthermore, more energy is required to remove an electron from an outer complete shell than an unfilled shell.

Crystal Structure

Now that all matter has been shown to consist of a fundamental unit called an atom, the arrangement of the atoms within a material can be investigated. Practically all of the inorganic (nonliving) solids occur in crystalline form. Even materials like iron, copper, and aluminum are crystalline in nature. A piece of iron is made of a great number of crystals lying in random positions throughout the material. A substance composed of a large number of crystals is called a *polycrystalline* material.

Crystal Lattice

If one were to examine common table salt under a magnifying glass, the small grains would appear as tiny cubes of salt. Each of these cubes has a precise atomic structure and constitutes a single crystal of salt. The arrangement of atoms in a salt (sodium chloride) crystal is shown in Fig. 11-2. In a salt crystal the atoms become ionized as the crystal is formed. The lines between the ions of sodium and chloride represent the chemical bonds which hold the crystal together. Due to the way in which the bonds form, every perfect crystal will be like every other crystal. This precise repeating arrangement of atoms within a crystal is called a crystal *lattice*. The

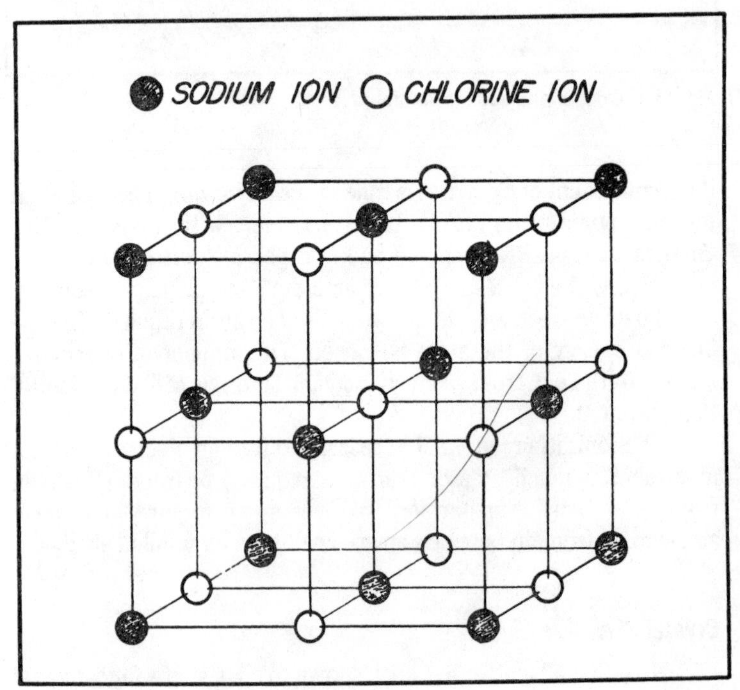

Fig. 11-2. Atomic lattice structure of salt.

physical properties of a material (hardness, tensile strength, etc.) are to a great degree dependent upon the lattice structure of the material.

Conductors, Semiconductors and Insulators

In the study of electronics, the association of matter and electricity is of paramount importance. Since every electronic device is constructed or parts made from ordinary matter, the effects of electricity on matter must be well understood. As a means of accomplishing this, all the elements of which matter is made can be placed into one of three categories: ***conductors, semiconductors,*** and ***insulators***. Conductors for example, are elements such as copper and silver which will conduct a flow of electricity very readily. Due to their good conducting abilities they are formed into wire and used whenever it is desired to transfer electrical energy from one point to another. Insulators (nonconductors) on the other hand, do not conduct electricity to any great degree and are therefore used when it is desirable to prevent a flow of electricity. Substances such as sulphur, rubber, and glass are good insulators. Materials such as germanium and silicon are not good conductors but cannot be used as insulators either, since their electrical characteristics fall between those of conductors and insulators. These in between materials which do not make good conductors, or good insulators, are classified as semiconductors.

The electrical conductivity of matter is ultimately dependent upon the energy levels of the atoms of which the material is constructed. In any solid material such as copper, the atoms which make up the molecular structure are bound together in the crystal lattice. Since the atoms of copper are firmly fixed in position within the lattice structure, they are not free to migrate through the material, and therefore cannot carry the electricity through the conductor.

Free Electrons

It has been shown earlier that by the process of ionization, electrons could be removed from the influence of the parent atom. These electrons, once removed from the atom, are capable of moving through the copper lattice under the influence of external forces. It is by virtue of the movement of these charged electrons that electrical energy is transported from place to place.

The ability of a material such as copper to conduct electricity must therefore depend on the number of dislodged electrons normally available within the lattice. Since copper is a good conductor, it must contain vast numbers of dislodged or free electrons.

To understand how the electrons become free, it is necessary to refer back to the electron energy levels within the atom. It was previously stated that if precisely the right amount of energy were added to an orbital electron, it would jump to a new orbit located farther from the nucleus. If the energy is sufficiently large, the jump can carry the electron to such a distance from the positive nucleus that the electron becomes free. Once free, the electron constitutes the charge carrier discussed above. The only problem remaining is to determine how the electron in the piece of copper obtains enough energy to become free.

After a moment's consideration you will realize that the average piece of copper contains some amount of heat energy. In fact, a piece of copper at room temperature (72 °F) is approximately 531 °F above absolute zero. This temperature indicates that the copper, although only warm to the touch, must contain a considerable amount of heat energy. The phonons of heat energy, along with other forms of natural radiation, elevate the electrons to the energy levels where they can become free.

Energy Gaps

From the preceding theories, one might wonder why all materials containing the same amount of heat energy do not conduct electricity equally well. The answer lies in the fact that the electrons in various materials require different amounts of energy to become free. This idea may be best developed by using energy level diagrams like the one in Fig. 11-3. In this model, the outer shell is depicted as having two energy bands called the valence bond band and the conduction band. Between these two energy bands is an energy gap called the forbidden gap, or forbidden band. Electrons residing in the lower band are considered to be firmly attached to the parent atoms and are not available for the conduction of electricity. In order for an electron to become a free electron, it must gain enough energy from external forces to jump the forbidden gap and appear in the conduction band. Once in the conduction band, the electron is free and can be made to move along through the conductor in the form of an electric current. The energy diagram for the insulator shows the insulator to have a very wide energy gap. This means that a large amount of energy must be added to each electron in an insulating material before it can become free. Thus at room temperature suffi-

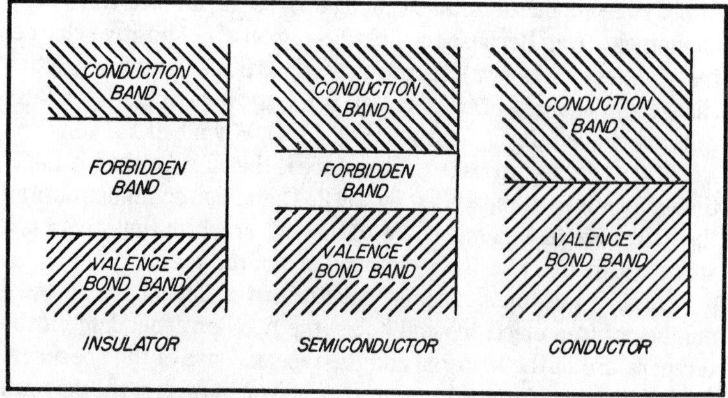

Fig. 11-3. Energy-level diagrams.

cient energy is not available to cause electrons to jump to the conduction band and the material has practically no free electrons. In comparing the energy-level diagrams for an insulator and a conductor, the conductor is seen to have little or no forbidden gap. Since this is true, under normal conditions the conduction band for a conductor contains a sufficient number of free electrons to make it a good conductor of electricity.

The semiconductor being neither a good conductor nor a good insulator has a energy gap which, on the energy level diagram, has a width between that of a conductor and that of an insulator.

In the following discussion the role of the conductor, semiconductor, and insulator will assume greater and greater importance as solid-state devices are developed and discussed. In fact, in the final analysis, all electronic phenomena are based on the electrical nature of matter.

Two Kinds of Current

The degree of difficulty in dislodging valence electrons from the nucleus of an atom determines whether the element is a conductor, semiconductor, or an insulator. When an electron is freed in a block of pure semiconductor material, it creates a hole which acts as a positively charged current carrier. Thus, electron liberation creates two currents, known as electron current and hole current.

Holes and electrons do not necessarily travel at the same rate, and when an electric field is applied, they are accelerated in opposite directions. The life spans (time until recombination) of a hole and a free electron in a given semiconductor sample are not necessarily the same. Hole conduction can be thought of as the unfilled tracks of a moving electron. Because the hole is a region of net positive charge, the apparent motion is like the flow of particles having a positive charge. An analogy of hole motion is the movement of the hole as balls are moved through a tube (Fig. 11-4). When ball 1 is removed from the tube, a space is left. This space is then filled by ball 2. Ball 3 then moves into the space left by ball 2. This action continues until all the balls have moved one space to the left, at which time there is a space left by ball 8 at the right-hand end of the tube.

A pure specimen of semiconductor material will have an equal number of free electrons and holes, the number depending on the temperature of the material and the type and size of the specimen. Such a specimen is called an *intrinsic semiconductor*; the current, which is borne equally by hole conduction and electron conduction, is called *intrinsic conduction*.

If a suitable impurity is added to the semiconductor, the resulting mixture can be made to have either an excess of electrons, thus causing more electron current, or an excess of holes, thus causing more hole current. An impure specimen of semiconductor material is known as an *extrinsic semiconductor*.

Impurity Donors and Acceptors

In the pure form, semiconductor materials are of little use in electronics. When a certain amount of impurity is added, however,

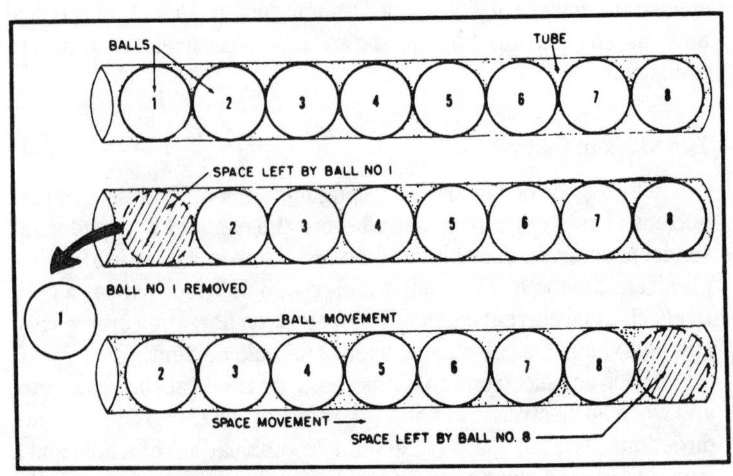

Fig. 11-4. Analogy of hole movement.

the material will have more (or fewer) free electrons than holes depending upon the kind of impurity added. Both forms of conduction will be present, but the majority carrier will be dominant. The holes are called positive carriers, and the electrons negative carriers. The one present in the greatest quantity is called the majority carrier; the other is called the minority carrier. The quality and quantity of the impurity are carefully controlled by a process known as doping. The added impurities will create either an excess or a deficiency of electrons, depending upon the kind of impurity added.

The impurities that are important in semiconductor materials are those impurities that align themselves in the regular lattice structure whether they have one valence electron too many, or one valence electron too few. The first type loses its extra electron easily and in so doing increases the conductivity of the material by contributing a free electron. This type of impurity has five valence electrons and is called a *pentavalent impurity*. Arsenic, antimony, bismuth, and phosphorous are pentavalent impurities. Because these materials give up or donate one electron to the material they are called donor impurities.

The second type of impurity tends to compensate for its deficiency of one valence electron by acquiring an electron from its neighbor. Impurities of this type in the lattice structure have only three electrons and are called *trivalent impurities*. Aluminum, indium, gallium, and boron are trivalent impurities. Because these materials accept one electron from the material they are called *acceptor impurities*.

N-Type Germanium

When a pentavalent (donor) impurity like arsenic is added to germanium it will form covalent bonds with the germanium atoms. Figure 11-5A illustrates an arsenic atom (As) in a germanium lattice structure. The arsenic atom has five valence electrons in its outer shell but uses only four of them to form covalent bonds with the germanium atoms, leaving one electron relatively free in the crystal structure. Because this type of material conducts by electron movement it is called a negative-carrier type or n-type semiconductor. Pure germanium can be converted into an n-type semiconductor by doping it with a donor impurity consisting of any element containing five electrons in its outer shell. The amount of the impurity added is very small; it is of the order of one atom of impurity in ten million atoms of germanium.

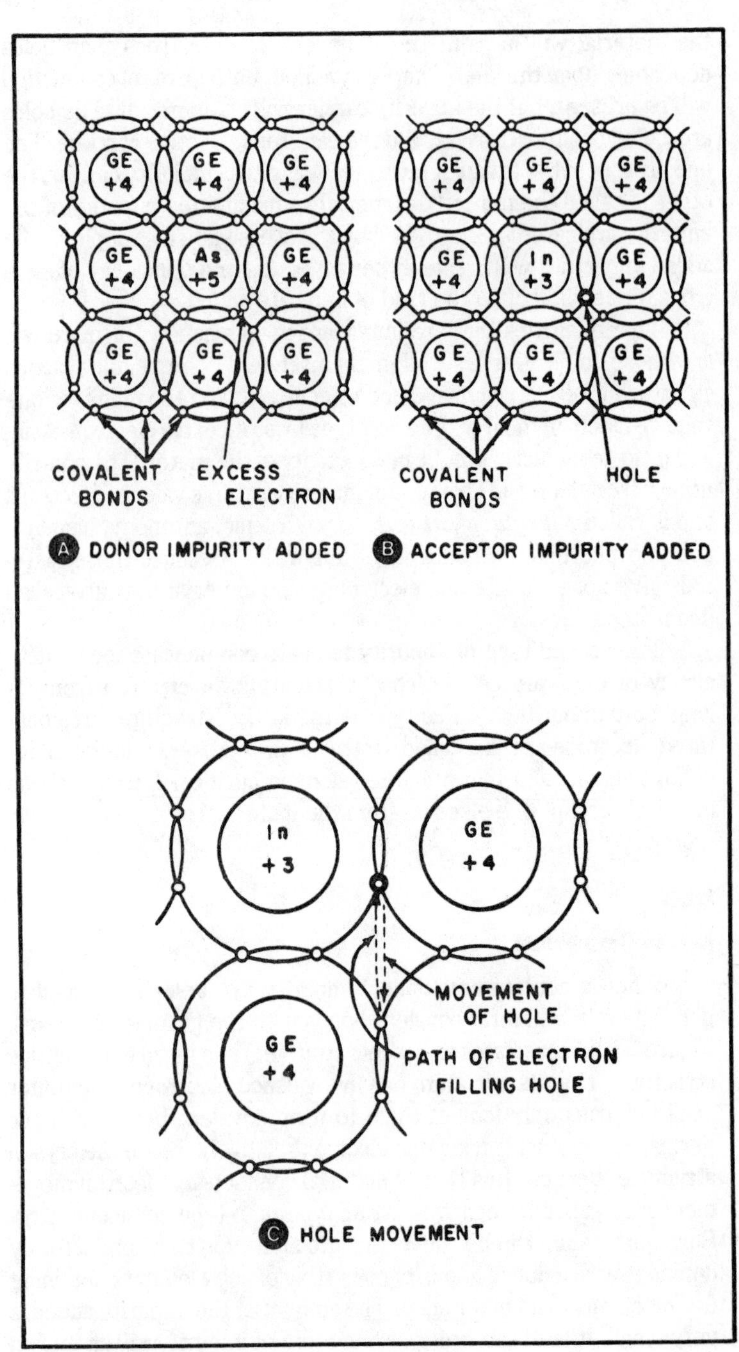

Fig. 11-5. Germanium lattice with impurities added.

P-Type Germanium

A trivalent (acceptor) impurity element can also be added to pure germanium to dope the material. In this case the impurity has one less electron than it needs to establish covalent bonds with four neighboring atoms. Thus in one covalent bond there will be only one electron instead of two. This arrangement leaves a hole in that covalent bond.

Figure 11-5B shows the germanium lattice structure with the addition of an indium atom (In). The indium atom has one electron less than it needs to form covalent bonds with the four neighboring atoms, thus creates a hole in the structure. Gallium and boron also exhibit these characteristics. The holes are present only if a trivalent impurity is used. Note that a hole carrier is not created by the removal of an electron from a neutral atom, but is created when a trivalent impurity enters into covalent bonds with a tetravalent (four valence electrons) crystal structure. Because this semiconductor material conducts by the movement of holes, which are positive charges, it is called a *positive-carrier type* or p-type semiconductor. When an electron fills a hole (Fig. 11-5C) the hole appears to move to the spot previously occupied by the electron.

PN Junction with Forward Bias

If you connected a battery across a pn germanium crystal with the positive terminal connected to the p-type germanium and the negative terminal to the n-type germanium, current would flow in proportion to the applied voltage. This type of connection is known as forward bias.

As shown in Fig. 11-6, the positive terminal of the battery repels the holes and causes them to move toward n-type germanium. Some of the holes enter the n-area.

The negative terminal of the battery repels the electrons and causes them to move toward the p-type germanium. And, in this case, some electrons enter the p-area.

Electrons and holes combine in a small area of diffusion on either side of the pn junction (between the dotted lines in the diagram). For each hole in the p-region that combines with an electron from the n-region, an electron from an electron-pair bond in the crystal near the positive terminal of the battery enters the battery at the positive terminal. This action creates a new hole, which moves toward the n-type germanium. For each electron that combines with a hole in the n-type material, an electron enters the crystal from the negative

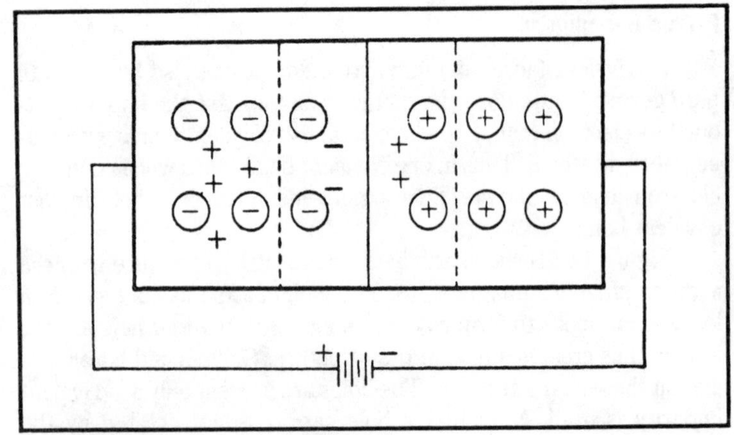

Fig. 11-6. A pn junction with forward bias.

terminal of the battery. The current flow in the p-region is mainly a flow of holes, and that in the n-region is mainly a flow of electrons.

INFRARED

Infrared, or IR as it is sometimes called, is assuming increasing importance in communications, mapping, missile and space vehicle guidance, and various military applications. Like radar, infrared technology was developed and used for military purposes during World War II. Unlike radar, however, the use of infrared has received little publicity. In many applications, infrared energy has advantages over radar and conventional radio. Infrared communications is the major use of injection lasers. Infrared, used as a medium of communication, is usually less susceptible to counterdetection and interference than is radar or visible light. This makes infrared lasers especially appealing to the military. Also, infrared is less subject to atmospheric absorption, which makes it appealing for communications with vehicles in space, from which signals are naturally weak. Infrared equipment is generally less expensive and complex than microwave equipment used for similar tasks.

INFRARED AND THE EM SPECTRUM

Infrared is a form of electromagnetic (EM) energy, with certain characteristics identical to those of light and rf waves. Among these characteristics, some of the more important ones (discussed in preceding chapters) are reflection, refraction, absorption, and speed of transmission. Infrared waves differ from light, rf, and other electro-

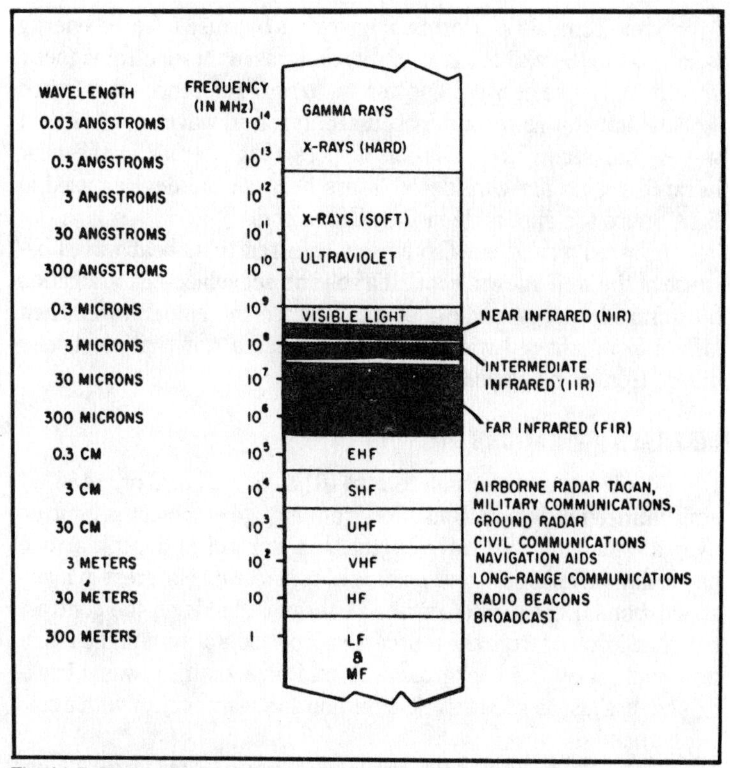

Fig. 11-7. Electromagnetic spectrum, showing different wavelength units used in different bands of the spectrum.

magnetic waves mainly in wavelength and frequency of oscillation.

The frequency of IR ranges from approximately 300,000 to 400 million megahertz. Its place in the electromagnetic spectrum (Fig. 11-7) is between visible light and the microwave region used for high-definition radars. This part of the spectrum is usually discussed in terms of wavelength, and it is generally accepted that the IR spectrum lies between 0.72 and 1000 microns.

The IR portion of the electromagnetic spectrum is frequently divided into the following three parts:

1. The near infrared (NIR), which extends from the visible region out to around 1.5 microns.
2. The intermediate infrared (IIR), which extends from 1.5 to 5.6 microns.
3. The far infrared (FIR), which extends from 5.6 microns to microwave frequencies.

Some confusion in terminology exists because infrared energy is so close to the visible range of wavelengths in the electromagnetic spectrum. Hence, it is not uncommon to hear references to infrared light, although this term is inaccurate. Infrared waves are not light waves, but stimulated emission devices that operate in the near infrared region are considered lasers because the devices used in light optics are applicable in the NIR region.

Infrared waves are also loosely referred to as heat waves because of the well-known fact that a solid object subjected to radiation by infrared energy undergoes an increase in temperature. Heat differs from infrared waves in much the same way that electricity differs from radio waves.

INFRARED RADIATIONS

All objects above absolute zero (0°K, or −273°C or −460°F) emit infrared radiation. Radiation from any given object is emitted over a wide range of wavelengths, but it reaches a peak at one particular wavelength. This point is of considerable interest in many applications. Detection of infrared energy depends on the contrast between infrared from the source under consideration and the radiation emitted by the background. A cold object with a warm background has just as good target definition as a warm object with a cold background.

There are a number of advantages in using IR for target detection rather than some other form of electromagnetic transmission. Infrared systems can be made smaller, lighter, less complex, and less expensive than other comparable systems. Further, infrared systems have higher target resolution.

The Marshall Space Flight Center of NASA developed an infrared laser-derived rendezvous guidance system for satellites. The system's purpose was to permit the linking up of vehicles in space. It used pulsed gallium-arsenide lasers on both the chase and target vehicles. The target vehicle, the one already in orbit and sought out by the chase vehicle, used a combination of multiplier phototube (discussed later in this chapter) and ordinary refraction optics to obtain a 10° field of view to permit long- and short-range tracking with gallium-arsenide light sources mounted on the chase vehicle. The target vehicle also had a gallium-arsenide laser transmitter to act as a beacon for the chase vehicle. Early on, NASA had considered ordinary microwave radar for the system, but conventional radar suffers from a lack of accuracy at very short distances (as during the final closure of the space vehicles) and poor ability to detect small amounts of movement, so the final choice was an infrared laser system.

EMISSIVITY

One of the most useful concepts employed when working with infrared is the "black body." A black body is defined as an object that absorbs all radiation incident on it. Conversely, the radiation emitted by a black body is the maximum possible for any given temperature. This black body, then, is a perfect absorber and radiator of infrared at all temperatures and for all wavelengths.

In order to compare the radiation emitted by an actual source with that of a perfect radiator, the concept of emissivity is employed. *Emissivity* is defined as the ratio of the total radiation emitted by any object at any temperature T to the total radiation that would be emitted by an ideal black body at the same temperature. The emissivity of any object depends on the amount of energy is surface can absorb. If the surface absorbs most of the infrared striking it, then it will emit a relatively high amount of radiation, and the emissivity of the object will be comparatively large. By the same reasoning, if the surface reflects most of the incident radiation, the object will have a relatively small emissivity. By definition, a black body has an emissivity of unity. Therefore, any other object will have an emissivity of less than one. The following information shows the emissivity of various surfaces.

Surface	Emissivity
Black body	1.00
Lampblack	0.95
Painted	0.90
Steel, cold rolled	0.60
Aluminum paint	0.25
Steel, stainless	0.09
Aluminum aircraft skin	0.08
Aluminum foil	0.04
Silvered mirror	0.02

Laws of Emissivity

The basic laws that describe the characteristics of infrared were developed first for black-body radiation, the ideal case, and then modified to describe the radiation from any source.

The most important parameter in determining the infrared characteristics of any body is the temperature of the body. As the temperature of an object changes, two specific changes in the infrared characteristics take place: the wavelength where peak radiation occurs is shifted, and the total energy radiated varies with the fourth power of the temperature. These fundamental relationships are given by the two laws that are discussed next.

Wien's Displacement Law

This law states that the wavelength at which maximum radiation occurs, $\lambda\mu$, is inversely proportional to the absolute temperature of the body. This can be expressed as

$$\lambda\mu = \frac{K}{T}$$

where the wavelength is given in microns, T is degrees of temperature in Kelvin, and the constant K has a value (for a black body) of about 2900. For example, a block of ice emits its peak energy at about 10 microns, but energy emitted from a jet engine peaks at about 3.5 microns. The relationship between the wavelength of peak radiation and the absolute temperature of a block body is given by Fig. 11-8.

Stefan-Boltzmann Law

This law states that the radiation intensity E is directly proportional to the fourth power of the absolute temperature, expressed as

$$E = \Sigma T^4$$

where E has dimensions of power per unit area, and Σ is the proportionality constant. In other words, if the temperature of an object is doubled, the radiation emitted from the object will be 16 times as much.

The Stefan-Boltzmann law can be modified to include the emissivity factor; by using the right units, the total radiation can be obtained by

$$E = \epsilon\Sigma T^4$$

where ϵ is the emissivity factor of the radiating surface.

A black body at an absolute temperature of 300°K (81°F) will radiate 46 milliwatts of power per square centimeter of its surface. A painted surface, such as the skin of a commercial airliner, at the same absolute temperature, will radiate 41 milliwatts per square centimeter. If this aluminum aircraft skin were not painted, the emissivity factor would be considerably smaller, and the radiation would be less than four milliwatts of power per square centimeter. As mentioned

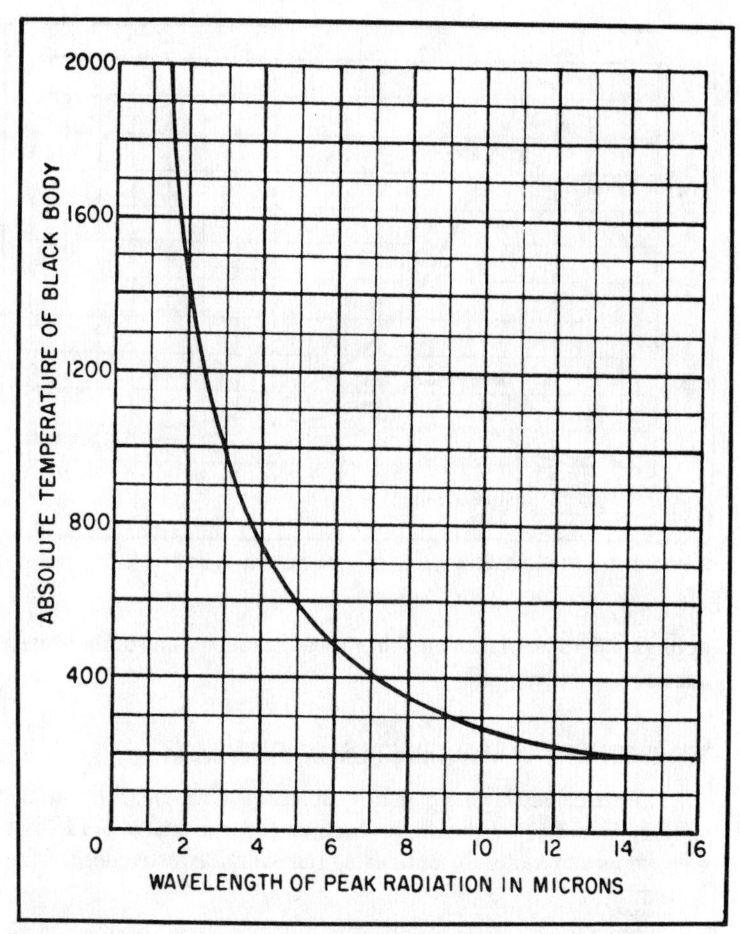

Fig. 11-8. The wavelength of the peak radiation from a black body in relation to its temperature.

previously, infrared from a source will be distributed over a good part of the spectrum, but the maximum radiation will be at some specific wavelength. Figure 11-9 shows the distribution of energy radiated from a black body at various temperatures.

For example, infrared from jet and rocket engine exhaust plumes is primarily due to the molecular excitation of water vapor and carbon dioxide, which are characteristic by-products of combustion. This molecular radiation peaks at 2.7 microns (due to water vapor and carbon dioxide) and 4.3 microns (due to carbon dioxide alone). In a practical situation, however, more usable radiation for

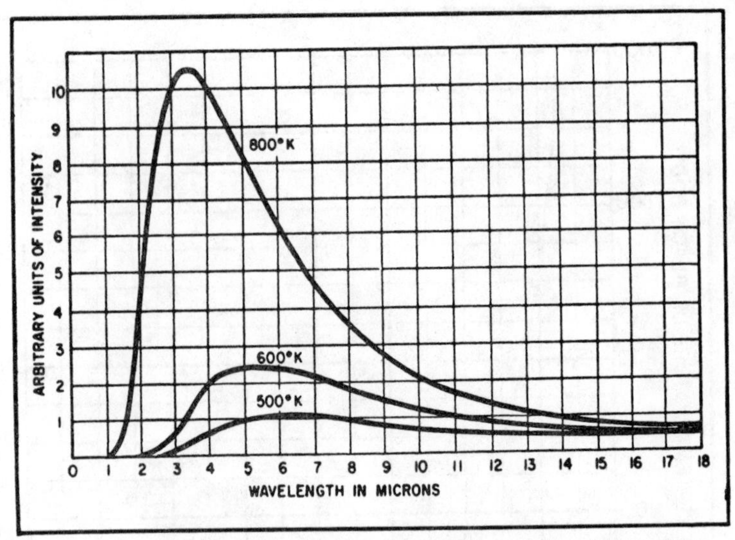

Fig. 11-9. The IR distribution curves for a black body at various temperatures.

detection purposes is obtained from the hot tailpipe and other heated surfaces.

TRANSMISSION CHARACTERISTICS OF INFRARED

The transmitting medium in most laser applications is the atmosphere. The effect of the atmosphere on the transmission of IR is a very important factor in considering the overall effectiveness of the system.

Atmospheric Absorption of Infrared

There are two primary sources of atmospheric attenuation; scattering by suspended solids and absorption by free molecules. These two influences on the attenuation are additive, but the absorption is the more important factor.

The amount of scattering caused by particles depends on the relationship between the wavelength of the radiated energy and the size of the particles. If the wavelength involved is considerably shorter than the dimensions of the particles, then the scattering is just about independent of wavelength. Since this is generally the case in the infrared spectrum, the attenuation due to scattering can be measured at one wavelength and then applied over a relatively wide band of wavelengths.

This technique does not work, however, with attenuation due to molecular absorption. The amount of absorption is closely connected with the wavelength. The two substances in the atmosphere that absorb the most radiation are water vapor and carbon dioxide. In both of these substances, there are several wavelength bands where the absorption is relatively large. This condition is caused by molecular resonance. (Each molecule has a natural frequency of vibration, or resonant frequency.)

The resonant frequencies of these molecules are in the infrared region, and their structure is such that this natural vibration creates an oscillation of electric charge in the molecules that increases the absorption. At low altitudes, where the atmosphere is relatively dense, this absorption is so great that in some wavelength bands the percentage of radiation transmitted drops rapidly to zero. Between these absorption bands are areas where the atmospheric attenuation is not so great. The transmission bands are called ***windows***. The transmission windows in the atmosphere are approximately as shown below.

0.95 – 1.05 microns
1.2 – 1.3 microns
1.5 – 1.8 microns
2.1 – 2.4 microns
3.0 – 5.0 microns
8.0 – 12.0 microns

Figure 11-10 shows the transmission characteristics of the atmosphere at two different altitudes. It will be noted by reference to Fig. 11-10 that the 2.7- and 4.3-micron radiation readings from a jet exhaust plume occur at frequencies of high absorption. The radiation resulting from hot tailpipe surfaces is a function of tailpipe material and temperature, usually peaking at about 3.5 microns. This frequency is in the region of an atmospheric window, or low-absorption band. It will also be noted that the absorption bands are much narrower at high altitudes due to the thinner atmosphere. These absorption bands will, therefore, be of lesser consideration in the design of high-altitude systems. Designers of laser communication systems take advantage of the windows in the infrared region.

OPTICAL DEVICES

Because of the similarity between infrared and visible light, optical devices are usually employed to gather and focus the infrared radiation. A simplified infrared optical arrangement is shown in Fig. 11-11. This sketch shows a simple optical system that could be used

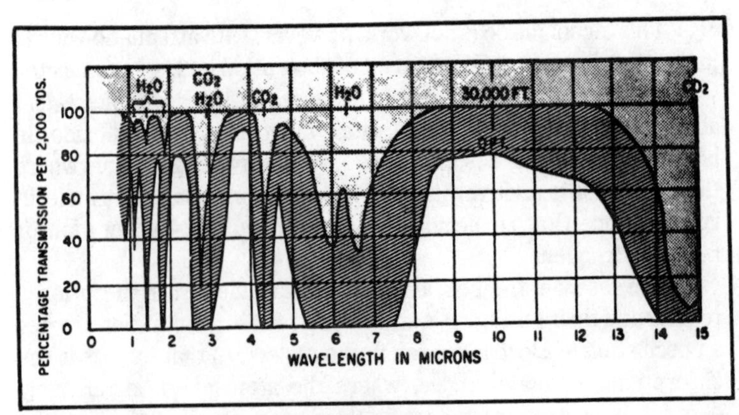

Fig. 11-10. Atmospheric transmission characteristics.

to focus the infrared radiation. The entire system is enclosed in a protective housing to shield the detector and optical system from the weather. The dome is a continuation of the protective housing and must be able to pass infrared radiation easily.

Typical materials from which domes are fabricated include glass, quartz, synthetic sapphires, germanium, and silicon. The transmission coefficient of the optical material is an important factor in the design of infrared equipment. For NIR, and generally for IIR, glass and quartz are satisfactory. Figure 11-12 shows that glass, quartz, and synthetic sapphires have excellent transmission charac-

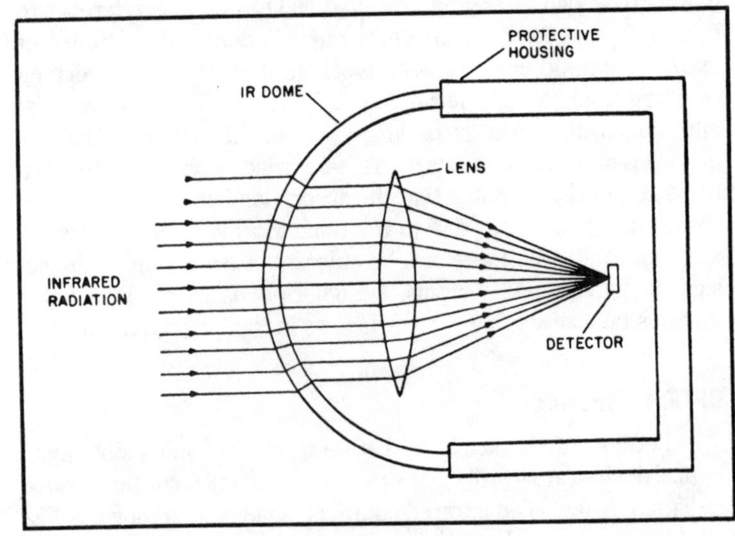

Fig. 11-11. Simple infrared optical arrangement.

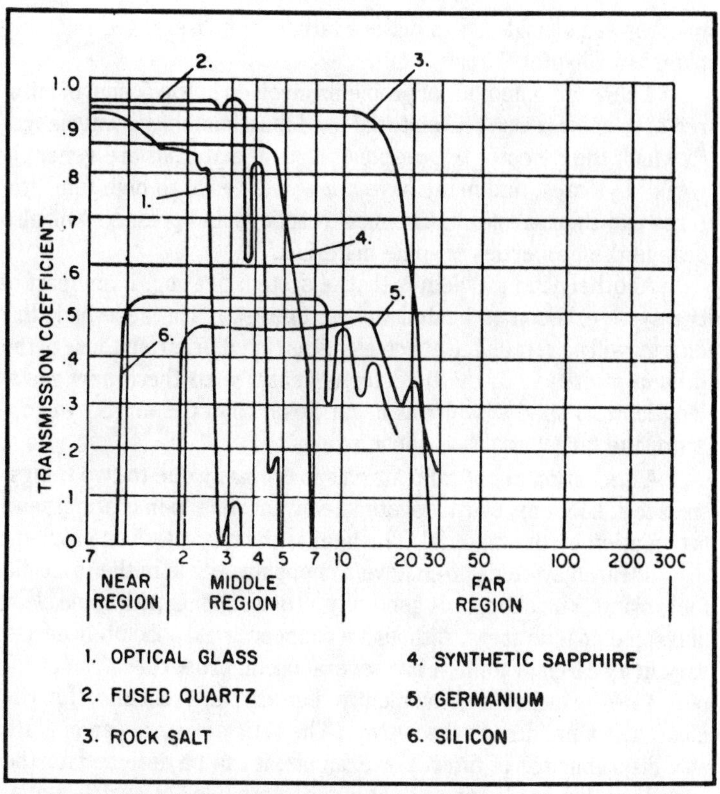

Fig. 11-12. Wavelength versus transmission coefficient.

teristics in the visible and near infrared regions. Optical glass is completely opaque to wavelengths longer than 3 microns. Quartz cuts off at 4 microns, and synthetic sapphire loses its transparency at wavelengths longer than 6 microns.

Germanium and silicon semiconductor materials are opaque to visible light and are transparent throughout most of the near and middle infrared regions.

A wholly different type of optics is required for FIR. Single crystals of silver chloride, rolled flat, are satisfactory windows for the transmission of FIR. Sodium chloride (rock salt), grown in single crystals and cut and ground into a lens or window, is excellent. Of course, rock salt is highly soluble in water and must be protected from the atmosphere's moisture. For FIR wavelengths longer than 15 microns, even these special optics lose their effectiveness. Stimulated emission devices operating at wavelengths longer than 15 microns are considered as masers rather than lasers, and the techniques in Chapter 6 apply.

There are a number of problems involved in the design of infrared optical systems. The material used must match the wavelength to which the detector will respond. Optical materials are generally weak physically, and many have a low resistance to high temperatures and thermal shock. Chemical reactions and pressure will also change the properties of some materials.

Another vital problem is that of system heating. If any part of the system becomes heated by the energy it has absorbed, this energy will be reradiated at wavelengths that differ from those of the original radiation. If the detector is sensitive to these new wavelengths, this source, which is much closer than the target, will obscure the target or cause ghost images.

Attenuation in optical materials is primarily due to two things: surface reflections, usually overcome by antireflection coatings, and attenuation by the material. The later is the more serious problem.

Infrared systems often have a chopping reticle in the principle focal plane. The chopper is generally a rotating disk with some clear and some opaque areas. Although a chopper is not absolutely necessary in a search system, it has several useful properties. The chopping rate furnishes a conveniently high carrier frequency for the electronic amplifiers in the system. The reticle pattern can operate as a discriminator or filter. For example, it can be designed for the type of backgrounds expected in order to provide better differentiation of target and background.

Optical filters are used in infrared instruments to isolate certain wavelength regions of interest, such as atmospheric windows, and to screen out undesired wavelengths. Three general types of filters are as follows:

1. Those that pass short waves.
2. Those that pass a particular band of waves.
3. Those that pass long waves.

INFRARED DETECTORS

One of the most critical components of an infrared system is the detector. The characteristics of the atmosphere and the source, if it is a military target, cannot be changed. Optical materials are somewhat standardized, as are display devices and control circuits. Although research and development in recent years has resulted in some very good all around detectors, it is still necessary to use care in choosing the proper one for a particular application.

Many variables must be considered in the choice of a detector. The characteristics of the radiation involved will determine what type of detector will best fit the needs of the system.

DETECTOR CHARACTERISTICS

In order to compare the relative merits of different detectors in different situations, it is necessary to introduce several parameters of detector operation. These parameters make it possible to discuss the characteristics of a particular detector in terms that can be applied to any detector.

Responsivity

When infrared energy strikes a detector material, either photoelectric or thermal, a change takes place that produces an electrical output signal, indicating that the material has been exposed to infrared. The amount of signal produced by each unit of input radiation intensity is called the responsivity R of the detector and is expressed as

$$R = \frac{\text{output}}{\text{input}}$$

generally in volts per watt. Many factors influence the responsivity, such as temperature of the detector and source, detector area, spectral distribution of the radiation, and detector time constant.

Spectral Response

One important influence on the responsivity of a detector is the change in detector sensitivity with changes in the wavelength of received radiation. The spectral limit of responsivity is considered to be the wavelength where the responsivity is half that of its maximum value. Since the spectral response is a nonlinear characteristic, it must be known for each wavelength considered, and any discussion of values must include details of the conditions involved.

Time Constant

In any infrared scanning system, the time constant of the detector must be such that the detector can fully respond before the radiation intensity changes. The time constant is considered to be the time required for the detector to develop 63% of its maximum output signal. The maximum scanning rate is dependent upon the time constant.

Noise

Noise exists in any circuit that carries current. Most outside noises can be eliminated or drastically reduced by shielding and proper design, but thermal noise is an ever-present problem.

Power supplies employed with infrared detectors must be extremely well filtered. Since the infrared radiation received by the detector is very small, noise of any appreciable amount could be sufficient to mask weak infrared signals or cause false targets. Transistors, for example, are useless in infrared amplifiers because of their inherent noise characteristics.

Many different types of noise are generated within infrared systems. Among these, the most important are as follows:

1. Current noise caused by bias currents within the detector.
2. Johnson (thermal) noise caused by thermal fluctuations in the detector material.

At low bias voltages, current noise is negligible and the output noise voltage consists almost entirely of Johnson noise. The current noise increases linearly with bias voltages and can eventually become the primary source of noise.

Noise Equivalent Power

Another useful and important detector parameter is the noise equivalent power (NEP) of a detector. The NEP is the radiation power in watts that must strike a detector to produce a signal response from the detector equal to the noise output (in other words, signal-to-noise ratio of one) over a reference bandwidth.

When comparing two different infrared detectors, the one with the lower NEP has the higher useful sensitivity. Since this tends to be confusing, another parameter is used—the detectivity. This figure is simply the reciprocal of the given NEP of a detector, so that the higher the detectivity a cell has, the higher its useful output. For example, a detector with an NEP of 4.0×10^{-9} will have a detectivity of 2.5×10^{8}.

The best infrared detector would be one having the greatest possible responsivity, the widest possible spectral response within the frequency band of interest, and the lowest possible NEP (that is, the highest possible detectivity). A properly chosen detector, with the necessary operating conditions, might have a maximum range of 90 miles with a signal-to-noise ratio of five from a one-square-meter target at 300°K. This range is equivalent to an ability to detect the infrared energy emitted by an ice cube at three miles.

PHOTODETECTORS

The three major types of photodetectors are photoconductive, photovoltaic, and photoemissive. The signal-to-noise ratio of the detector is the limiting factor in determining its effectiveness. These are used in the near infrared spectrum.

Photoconductors use a semiconductor crystal that absorbs the photon energy from the radiation that strikes the surface of the crystal, changing its resistance or conductivity. Quite a number of materials have been used for this type of photodetector, including lead sulphide, lead telluride, lead selenide, cadmium sulphide, and many others. Gold-doped germanium has been established as a good detector material, but some difficulties, such as the long time constant, must be considered.

Photovoltaic cells are similar to solar cells. These are semiconductors with a high-resistance photosensitive barrier between the two layers. When exposed to infrared, this cell builds up a potential difference between the two layers.

Photoemissive cells operate on the principle that electronic emission is caused by exposing the cathode to infrared. In this type of emission, the number of electrons emitted depends on the intensity of the infrared striking the cathode.

A phototube detector is shown in Fig. 11-13. When infrared waves strike the photocathode, they cause electrons to be emitted from its surface. These electrons are attracted to a positively charged anode in the center of the photocell, producing a current flow in the cathode-to-anode circuit. The amount of current produced varies in proportion to the amount of light striking the photocathode. The composition of the coating on the photocathode determines its sensitivity to a particular frequency. Photocells with different cathode coatings are used in the field of spectrum analysis. A single cell supplemented by a set of color filters can also be used for this purpose. If the characteristics of the radiation emitted by a target are known, a sensing element can be selected that is most sensitive to the radiation with the known characteristics. By using filters to exclude radiation with unwanted characteristics, the selectivity of the system can be increased to a higher degree.

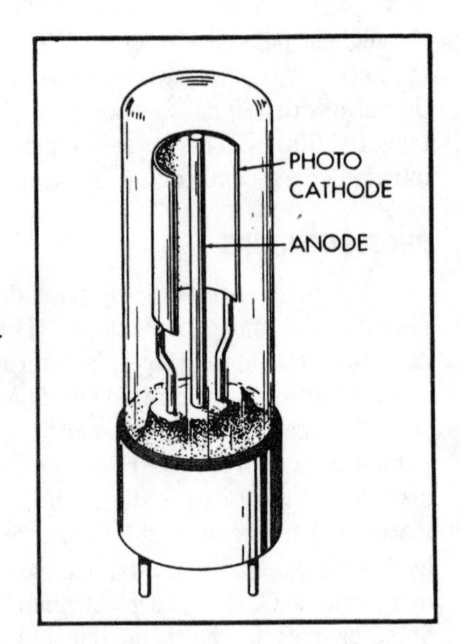

Fig. 11-13. Phototube.

For use with an infrared sensor, a filter can be selected that would absorb violet, blue, and yellow light. This would permit the infrared to pass through to the sensing element. Usually, no single filter screen will produce the desired result; so two or more filters may be necessary to remove all or most of the unwanted wavelengths. Filters must be selected that will not reduce the intensity of the desired form of radiation below the sensitivity range of the sensing unit.

The photomultiplier tube, shown in Fig. 11-14, is often used in infrared systems. The photoanodes are arranged so the electrons emitted by the cathode are attracted to a positively charged anode, from which an increased number of electrons is released by secondary emission. These electrons are attracted to the next anode, which is at a higher positive potential than the preceding anode. A greater number of electrons are released. This process continues throughout the photomultiplier tube until the final anode is reached, which can be at a very high positive potential. In this manner, a very high current flow can be produced from an initially small amount of photoemission. The use of photomultiplier tubes is restricted to systems in which high anode voltages are available. Photomultiplier tubes are highly efficient as infrared-sensing devices because of their sensitivity and amplification.

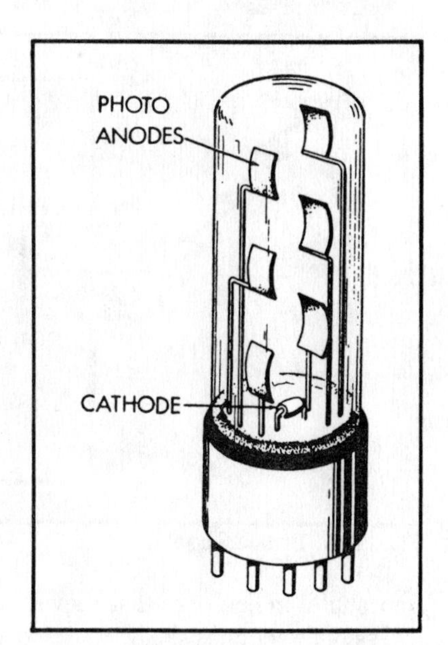

Fig. 11-14. Photomultiplier tube.

THERMAL DETECTORS

Thermal detection is accomplished by sensing the change in temperature of the detector material as a result of infrared striking its surface. There are three different types of sensing elements employed in modern detectors to sense this temperature change. These are as follows:

1. A *thermopile* is a series combination of several thermocouples.
2. A *bolometer* senses the change in resistance of the detector material.
3. A *pneumatic cell* uses the expansion of a gas as an indicator.

Thermocouple

One of the basic heat detectors is the thermocouple. When two dissimilar metals such as iron and copper are joined together and heat is applied to the junction, a measurable voltage is generated between them. Figure 11-15 shows a basic thermocouple.

The voltage difference between the two metals is quite small, but the sensitivity can be increased to a point at which the thermocouple becomes useful as a detector of infrared. The increase in sensitivity is obtained by connecting or stacking a number of ther-

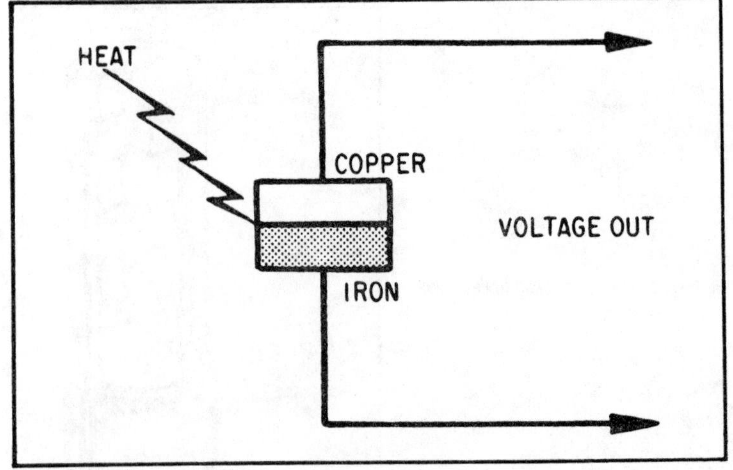

Fig. 11-15. Thermocouple.

mocouples in series so that they form a thermopile. The complete thermopile action is similar to that obtained when a number of flashlight cells is connected in series; that is, the output of each individual thermocouple is added to the output of the others. Thus ten thermocouples, with individual output voltages of 0.001 volt, would have a total output of 0.01 volt when connected in series.

The effective sensitivity of a thermopile can be further increased by mounting it at the focal point of a parabolic reflector. When this method is used, infrared given off by the target is focused on the thermopile by the reflector.

Bolometers

A bolometer is a very sensitive material whose resistance will vary, depending on the amount of infrared to which it is exposed. There are two main classes of bolometers: barretters and thermistors.

A barretter consists of a short length of very fine wire, usually platinum, which has a positive temperature coefficient. That is, its resistance increases with temperature. (A negative coefficient would mean that the resistance decreases with increasing temperature.)

The thermistor is a type of variable resistor made of semiconductor material, such as oxides of manganese, nickel, cobalt, selenium, or copper. The thermistor has a negative temperature coefficient of resistance. Thermistors are made in the form of beads, disks, rods, and flakes, some of which are shown in Fig. 11-16.

The heat-sensitive materials of thermistors are mixed in various proportions to provide the specific characteristics of resistance versus temperature necessary for target detection.

Figure 11-17 shows the change in resistance that can be produced in a typical thermistor material and in a barretter. This comparison shows that the thermistor has the steeper temperature-coefficient-of-resistance curve; it is therefore the more sensitive of the two sensors.

One simple type of infrared detector consists of two thin strips of platinum that are used to form two arms of a Wheatstone bridge. To increase the thermal sensitivity of the strips, each is blackened on one side. The infrared to be measured is applied to one of the strips and is absorbed by its blackened surface. As the strip absorbs heat, its resistance changes and unbalances the bridge. The imbalance causes a change in current produced by an external voltage applied to the input terminals of the bridge.

A device of this kind is shown in Fig. 11-18. It consists of four nickel strips supported by mounting bars, which have electrical leads attached to them. A silvered parabolic reflector (mirror) is used to

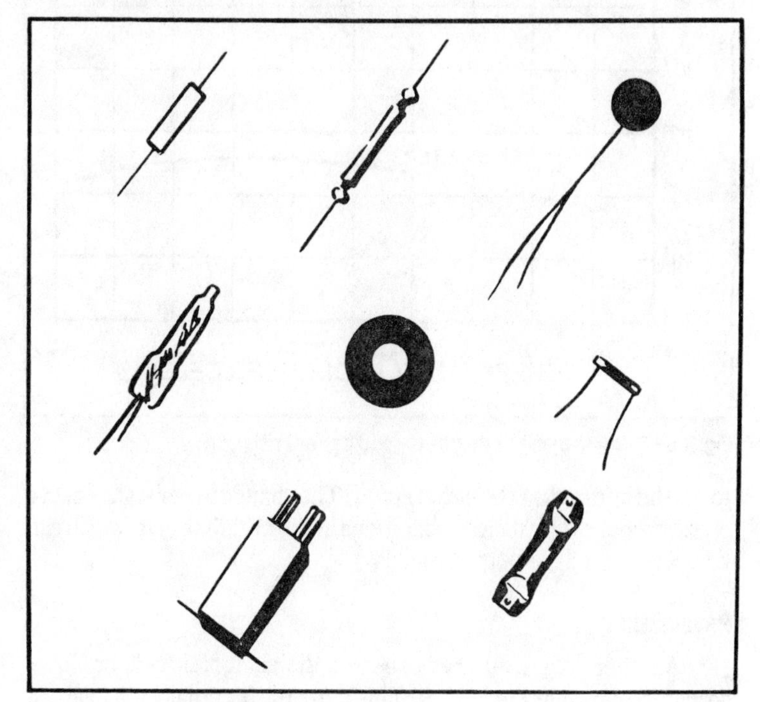

Fig. 11-16. Various thermistors.

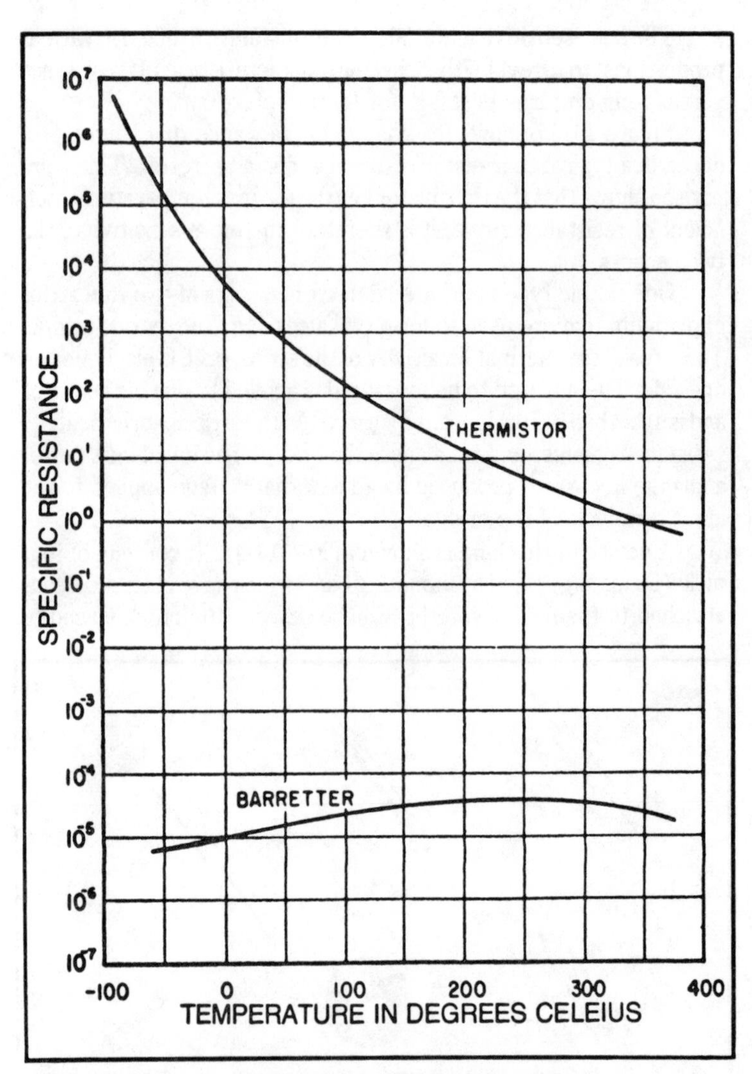

Fig. 11-17. Comparison of thermistor and barretter sensitivity.

focus the infrared on the nickel strips. The change in resistance of the strips produces unbalanced conditions in the bridge circuit, which can be used to produce output signals.

Pneumatic Cell

Another unique infrared sensor is the pneumatic cell, or Golay detector, shown in Fig. 11-19. This detector is actually a miniature heat engine.

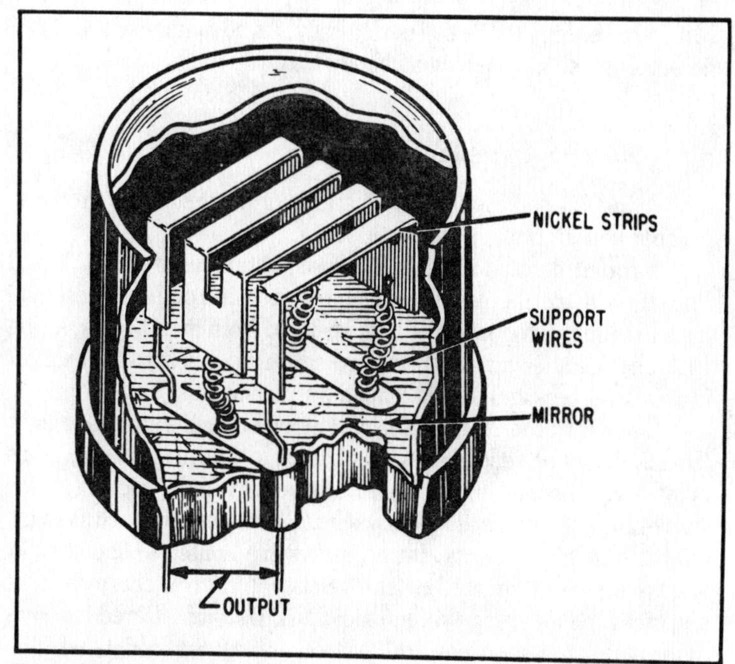

Fig. 11-18. Infrared detecting device.

Infrared energy entering the window causes expansion of a volume of gas located between the reflecting diaphragm and the window. The lamp at the other end of the detector emits a light beam that is focused by the lens and then passes through the grid and onto the reflecting diaphragm. The expansion and contraction of the gases between the window and diaphragm will cause the diaphragm to change its shape in response to changes in the amount of infrared energy entering the window.

Changes in the shape of the diaphragm will cause its light-reflective properties to vary accordingly, modulating its light output. The light reflected from the diaphragm will then pass back through the grid, which is designed to intensify the variations of the reflected light. After passing through the grid, some of the light strikes the mirror and is reflected to a photocell of high sensitivity (not shown in the figure). The modulated output of the photocell is a voltage proportional to the intensity of the infrared entering the window.

The Golay detector has the most rapid response of any infrared detector, but it can operate only when the radiant energy is received intermittently. An optical chopper can be used to interrupt the flow of infrared energy to the cell periodically. An advantage of the Golay detector is its extremely wide bandwidth.

INFRARED APPLICATIONS

Industrial and military applications of infrared have grown in number considerably in the last decade.

Infrared devices found their first military uses during World War II as a snooperscope, which operated in total darkness and outlined enemy troops by the heat radiated from their bodies. A rifle with a device known as a sniperscope made it possible to see a target in total darkness and fire on it with normal accuracy.

Since infrared is invisible but behaves much like visible light (that is, it can be reflected and controlled in a beam pattern), it has proved very useful for communications. Infrared, when used for short-range communication between sea-level stations such as ships, affords excellent security. Infrared rays are limited to line of sight and are rapidly attenuated at sea level; hence infrared acts in favor of security for short-range communications. Uses of infrared for communications require a powerful source that can be modulated. The infrared laser, discussed in the next chapter, provides such a source.

Infrared laser photography can produce very high-resolution pictures. Night photography using infrared laser photography can produce a better visual presentation of terrain than the best mapping radar.

Infrared has also invaded the navigation field. Ground speed indicators are being developed that will compete with Doppler radar. Anticollision circuits are also under experimentation. Infrared radar uses a pulsed infrared source and receives reflected infrared energy as in microwave radar.

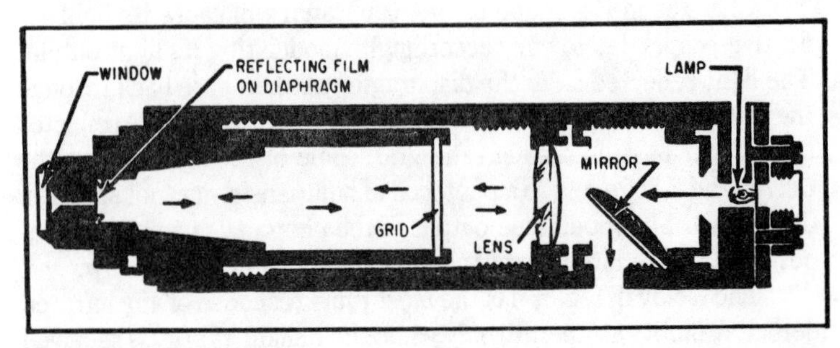

Fig. 11-19. Golay detector.

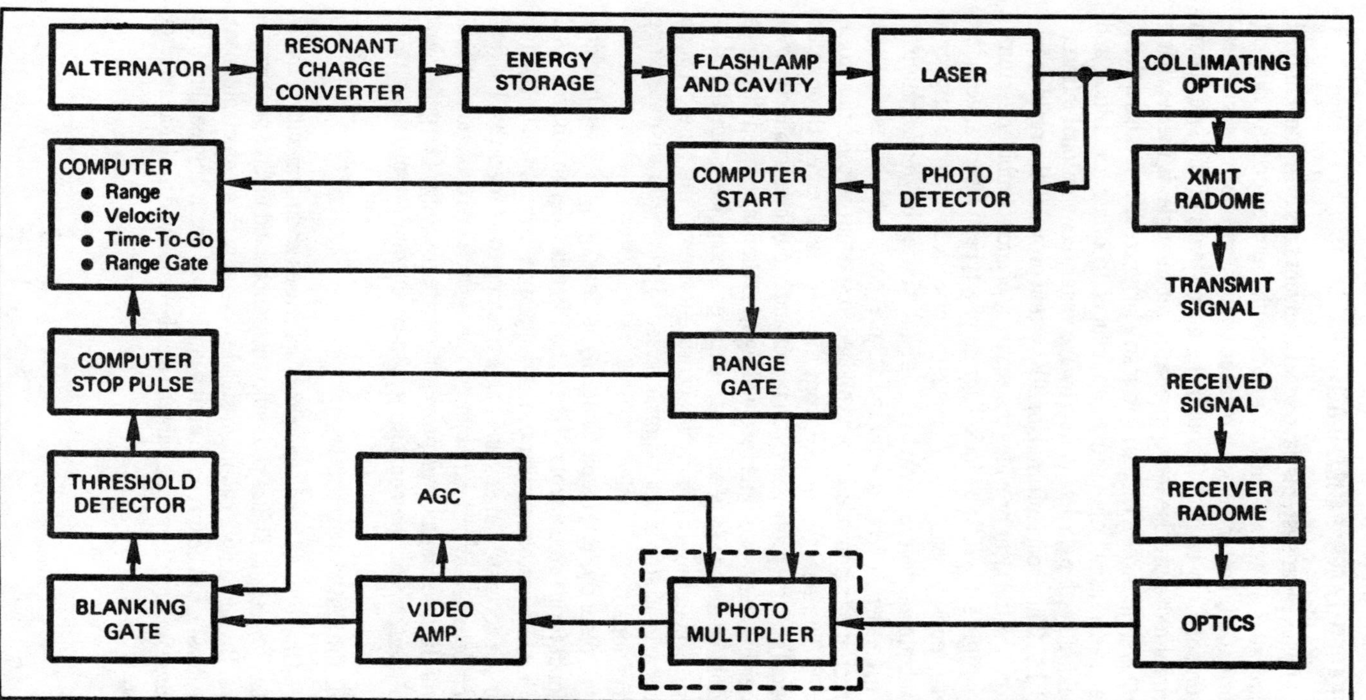

Fig. 11-20. Laser rangefinder. (Courtesy of Raytheon Company).

INFRARED LASER RANGEFINDER

Laser radar systems have several advantages over conventional radio radar systems. The laser radars give more accurate ranging information, better velocity measurements, and better angle measurement (to a target). Though the narrowness of the laser beam permits more precise location of a target in angle, it means that a laser system takes a long time to scan a large area. This means that a laser radar usually is used as an adjunct to other acquisition and tracking systems, such as microwave radar, conventional infrared, and TV. The radar, infrared, or TV system provides the rapid acquisition, and the laser provides the accuracy of such combined systems.

Figure 11-20 is a block diagram of a typical laser rangefinder. Here, energy stored in a charging network is used to trigger a flashlamp inside a laser cavity. The light emitted by the flashlamp, in turn, causes the laser to emit radiation. A portion of the laser energy detected by a photodetector is utilized to start an elapsed-time computer. The energy emitted from the laser is directed through an optical system, propagated to a target and back, then is collected by a receiving telescope. The received signal is then processed through a suitable detector to generate a computer stop pulse. Since the computer's counting capability is accurately known, the time between the start and stop of laser signals is used to provide an accurate range indication.

Presently, Raytheon Company is developing a laser tracking adjunct for a fire-control radar system. In this role, an infrared YAG laser radar system, similar to the one just described, will be used to provide accurate ranging and angle tracking to a target that has first been detected by ordinary microwave radar. The resolution capability of the laser adjunct will provide accurate tracking of targets at low altitudes, where the conventional radar encounters problems.

THE CADMIUM-SULFIDE

One type of semiconductor laser uses cadmium-sulfide as its active material. This is the same substance used in CdS photoresistors. By pumping a cadmium-sulfide crystal with an electron beam, the crystal can be forced to emit a fan-shaped beam of green light. Of course, many other materials can also be employed in semiconductor lasers.

12

Injection Lasers: Theory and Experiments

The newest and smallest member of the family of lasers is a tiny block of semiconductor crystal called the injection laser. This represents a giant step toward practical realization of the great potential in laser light. The injection laser is closely related to the optically pumped laser crystal. In 1960, the optically pumped crystal gave a new form of light to science and industry. For the first time, light possessed the useful properties of radio waves and microwaves; it retained its phase for a large number of cycles.

As with radio waves, light consists of electromagnetic oscillations propagated through space. Prior to the laser, light waves had not been produced with the spectral purity that characterizes radio waves. The laser changed this situation by producing light that is extremely well-defined in frequency, and spatially uniform.

The most striking feature of the injection laser is the simplicity with which its electrons are pumped, or excited, into high-energy states. Ultimately, the energy used to pump almost all lasers is electrical, but this electrical energy must be converted into energy of excited electrons in a laser.

The simplicity of injection laser pumping can be seen easily by comparing it to two other types (see Fig. 12-1). In the case of the optically pumped laser, electrical current operates a lamp that emits radiation in the pumping band of the laser. Radiation in the pumping band produces excited atoms. Similarly, in the gas laser the electrical

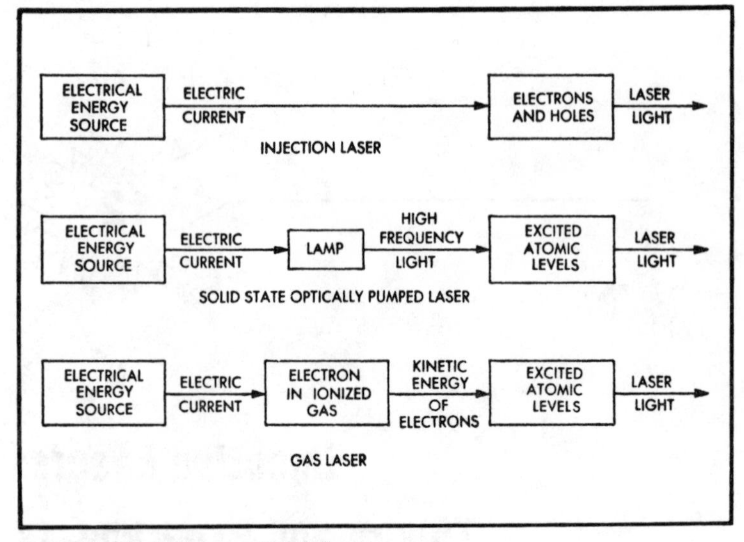

Fig. 12-1. Block diagram of basic lasers.

energy is converted first into kinetic energy of electrons. The high-energy electrons then excite atoms of the gas by colliding with them. Injection lasers use electrical energy directly, pumping electrons to high energy states by injecting electrons and holes across a pn junction.

Eliminating an intermediate stage in the transformation of electrical energy to light has two important advantages. First, whereas the conversion of energy from one form to another is usually inefficient, the injection laser's direct transformation process brings the efficiency to a relatively high level. Efficiency of the HeNe laser is typically as follows: injection type, 20%, optically pumped, 1%; and gas, 0.01%. A CO_2 type is typically 15% efficient and CO has been reported to be in excess of 40%.

The second advantage is the simplified equipment used in pumping electrons directly with electrical energy. Lasers that have to be pumped through an intermediate form of energy need bulky, expensive equipment for the transformation. The injection laser eliminates this auxiliary apparatus. This second advantage of the injection laser is especially apparent when compared with the optically pumped laser, which is itself a small crystal of fluorescent material. This small crystal must be surrounded by a much larger array of lamps and reflectors when in operation.

INJECTION LASERS AND LEDS

The injection laser is very closely related to the common LED, or light-emitting diode. When a voltage above a certain threshold level is applied to a LED, it glows.

In injection lasers there is a second critical threshold point, called J_{Th} or I_{Th}. When the applied current is below this threshold, the device functions exactly like an ordinary LED. That is, the device glows with a relatively broad spectrum of wavelengths. The light is emitted in a wide pattern of radiation.

But if the applied current to an injection laser exceeds the threshold level, the emitted light thins down to a narrow beam. The emitted laser beam escapes from both of the laser's end facets, unless one is coated with a reflective (usually gold) film.

To show just how closely related the injection laser and the LED are, the physical structure of a typical LED is illustrated in Fig. 12-2, while Fig. 12-3 shows the physical structure (simplified) of a diffused junction laser made from a similar semiconductor wafer. The way these devices function is illustrated in Fig. 12-4.

EARLY DEVICES

While injection lasers are relatively new, there has been considerable progress since they were first developed.

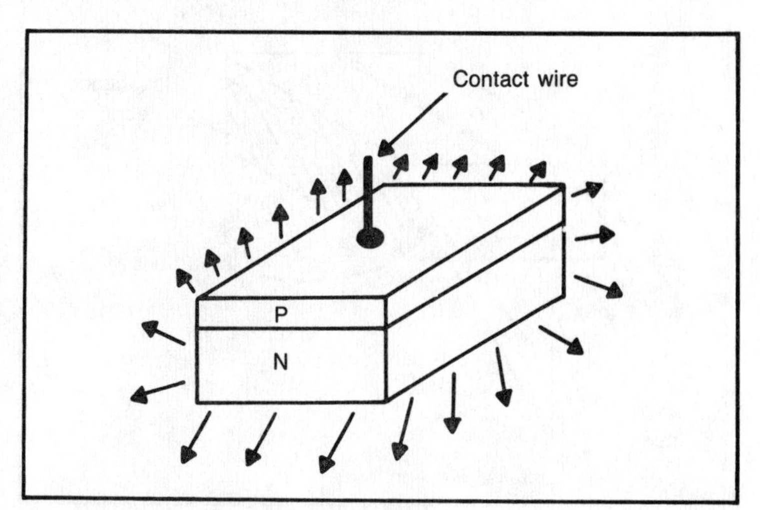

Fig. 12-2. Physical structure of a typical LED.

The phenomenon of pn junction electroluminescence was first noted in 1923. It was found that light was often emitted when point electrodes were placed on certain silicon carbide crystals and current passed through them. Modern semiconductor theory explains this phenomenon. If electrons are injected into p-type material, they combine spontaneously with the majority carriers existing in the material, and can emit radiation in the recombination process.

Previously, pn junction electroluminescence had never been a technologically useful phenomenon because its efficiency was low. Most electrons do not emit a photon in recombining with a hole, but rather convert combination energy to heat. The function of junction electroluminescence in science and technology was altered rapidly in the early 1960s.

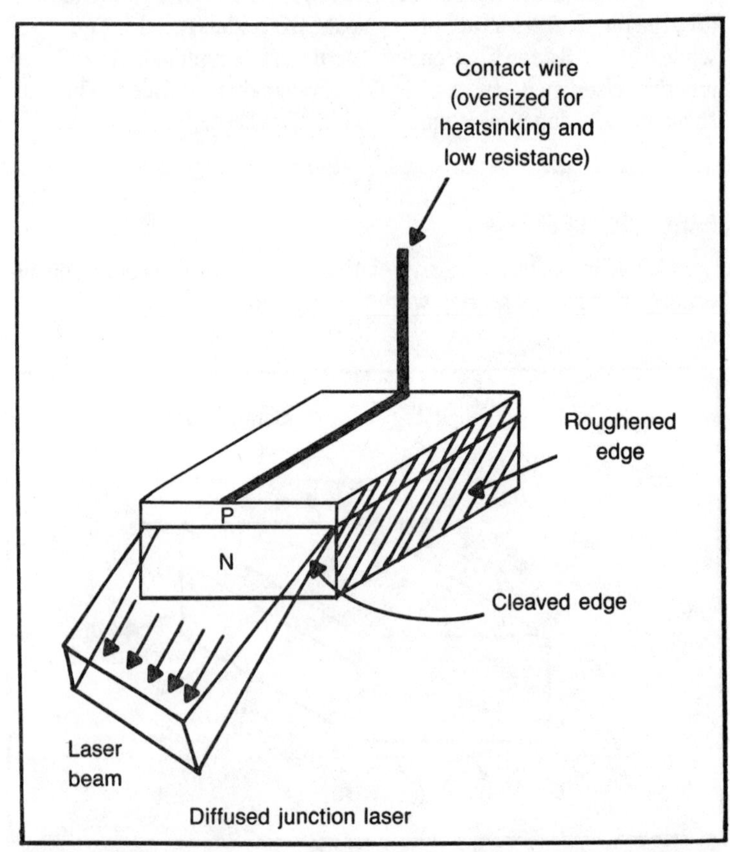

Fig. 12-3. Physical structure of a diffused-junction laser made from a semiconductor wafer.

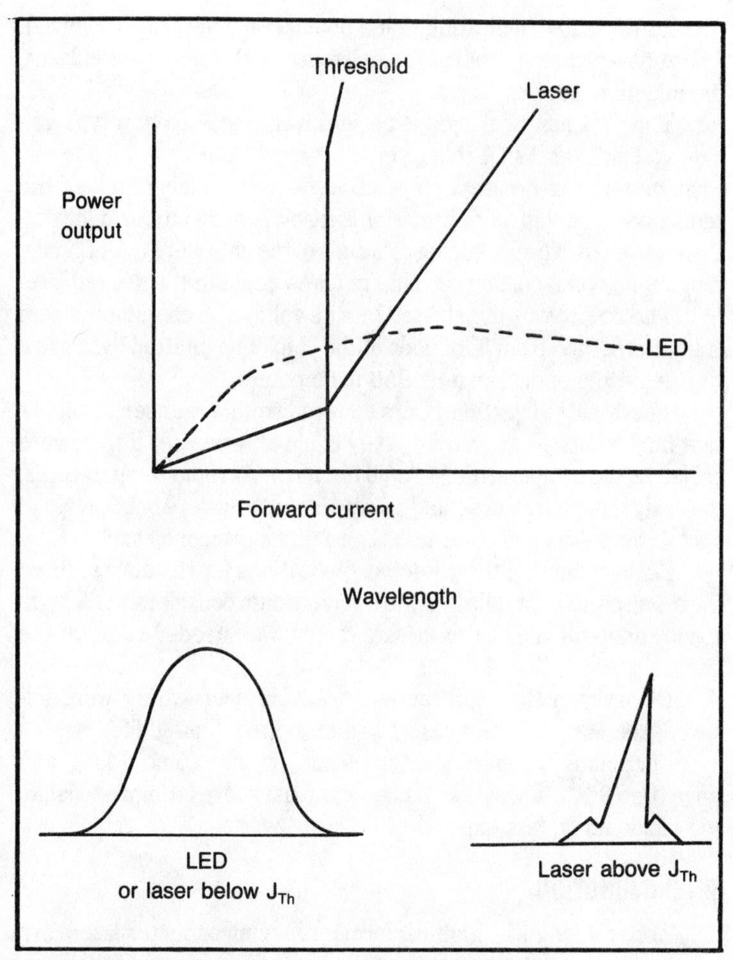

Fig. 12-4. How the devices of Figs. 12-2 and 12-3 function.

In 1960, gas lasers and optically pumped solid lasers stimulated research into the possibility of a semiconductor junction laser. The discovery in 1961 and 1962 that the efficiency of combination radiation was very high in gallium arsenide focused attention on this compound as a potential laser material. In gallium arsenide, most of the electrons emit a photon during the combination, instead of energy as heat through some nonradiative process.

Laser action is observed when large forward currents are applied to special gallium arsenide diodes maintained at very low (cryogenic) temperatures. The start of laser action occurs at a definite current threshold, indicating that a population inversion is involved. These observations confirm the existence of this new type of laser, the injection laser.

The first practical injection lasers were made from n-type gallium arsenide (GaAs). A thin layer of p-type dopant was added to this basic material to create a pn junction. The two facing ends of the semiconductor chip were precision cut and polished to form mirror-like end facets. The remaining two ends of the chip were intentionally roughened when cutting the chip of GaAs down to the desired size.

The device was then biased by a dc voltage. A continuous beam of light emerged from both ends of the chip. The emitted light had a wavelength of approximately 850 nanometers.

These early injection lasers suffered from a number of significant limitations. For one thing, they could only be used if they were cooled to the temperature of liquid nitrogen. At room temperatures, the early injection lasers could safely put out only very brief pulses of light. The pulses had to be less than 200 nanoseconds each.

Besides limiting the potential applications for the device, these short pulses also complicated the circuit design considerably. A high-speed, miniaturized, pulse-driving circuit was needed to drive the laser.

Even when these precautions to avoid overheating were followed, the early injection lasers had extremely limited lifetimes.

Fortunately, semiconductor technology has come a long way since the 1960s. Today's semiconductor lasers are far more reliable, and incidentally, less expensive.

HETEROJUNCTIONS

Today's long-life, high-performance, semiconductor lasers are largely due to the development of the heterojunction. A ***heterojunction*** is a junction of a pair of differing semiconductor materials. For example, a heterojunction can be made from gallium arsenide (GaAs) and aluminum gallium arsenide (AlGaAs).

A heterojunction like this sandwiches the light-emitting pn junction between two or more semiconductor layers. This lowers the laser's threshold and improves its efficiency because the generation and emergence of the emitted light is restricted to the junction region.

This light confinement by the heterojunction is a result of the different refractive indexes of the two joined semiconductor materials. To work properly, the refractive index of the main semiconductor (in which the pn junction is formed) is higher than the border semiconductor.

A heterojunction essentially functions as a mirror to the light waves trying to travel between them. This process is often called *wave guiding*, because the light waves are "guided" to where we want them to be. A similar phenomenon is at the heart of fiber optics. Light passes through a core of plastic or glass fibers, even if the fibers are bent at angles. The core fibers are surrounded by an outer coat of plastic or glass having a lower refractive index than the core fibers, thus preventing the light from escaping. The light waves are forcibly "guided" to the desired destination.

There are a great many different types of heterojunction lasers. Most can be divided into two important classes—the single-heterojunction (SH) laser, and the double-heterojunction (DH) laser. The basic physical difference between these two types is rather self-evident from their names, but there are some very important functional differences.

Single-Heterojunction Lasers

Single-heterojunction (SH) lasers are characterized by high output power. Some of these devices can put out as much as 50 watts with each pulse. However, this high output power comes at a price. SH laser diodes require considerably higher operating current than double-heterojunction lasers. A driver circuit capable of putting out clean, short-duration pulses with extremely fast rise-times must be used to power a SH laser. In most designs the current is gated, or a capacitor is discharged through a device such as an avalanche transistor, a VFET, an SCR, or a four-layer diode. A typical SH-laser driver circuit is shown in Fig. 12-5.

As the name indicates, a single-heterojunction laser has a heterojunction on just one side of the light-emitting pn junction.

The size of the heterojunction is directly linked to the output power. As a rule of thumb, for every mil (0.001 inch) of junction width, a full watt of output power may be obtained.

The smallest SH laser diodes are about three mils wide, and will thus emit about three to four watts at ten amps of forward current. The nominal threshold current is usually somewhere around four amps.

The chief advantage of the SH laser is that it is much more efficient than a diffused-junction laser. The chief disadvantage is that an SH laser cannot be operated continuously at room temperature.

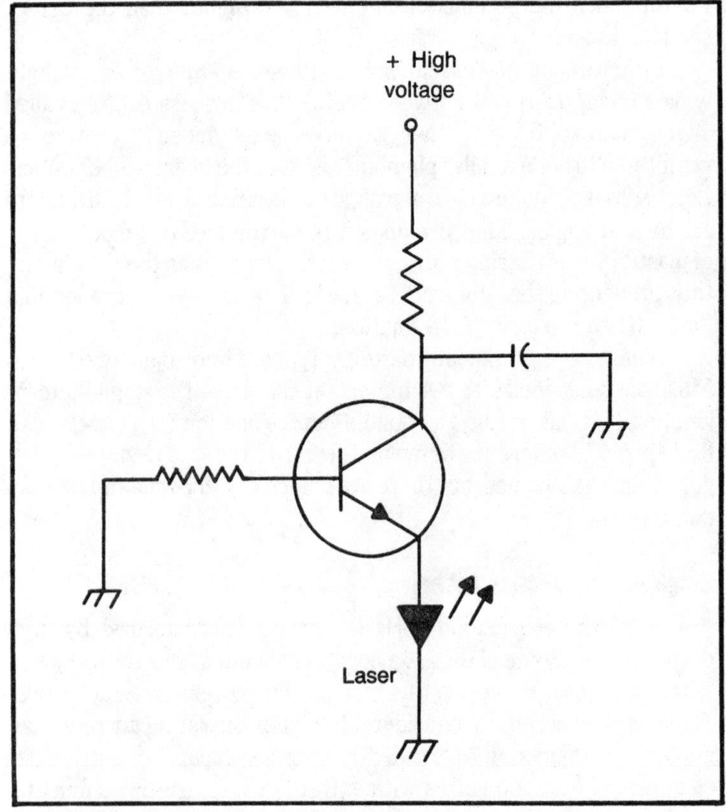

Fig. 12-5. A typical SH laser driver circuit.

At room temperature an SH laser must be driven by fast rise-time pulses of no more than 200 nanoseconds each.

Because of its high power output per pulse, the SH laser is well suited for applications like long-distance pulse-modulated communications and range-finding.

Double-Heterojunction Lasers

A double-heterojunction (DH) laser features a heterojunction on either side of the light-emitting pn junction. There are actually a number of different types of DH lasers.

Unlike the SH laser just discussed, the DH laser can be operated continuously at room temperature. The heterojunctions on either side of the pn junction efficiently confine the emitted light to an extremely thin area which runs along the length of the junction.

Unfortunately, the power supply requirements for the DH laser are complicated by the fact that the operating current is very temperature sensitive. If the temperature shifts just a few degrees, the laser's threshold will be altered significantly — possibly even enough to stop the lasing action, or even to destroy the laser diode. Clearly, a DH laser must be held at a constant temperature, or the supply for the forward current must include a thermal-tracking circuit for temperature compensation. Ideally, both of these precautions should be taken.

Often a special process known as stripe-geometry is employed in the production of DH laser diodes. In a *stripe-geometry* device everything except a narrow stripe of one electrode is insulated from the upper surface of the laser chip. This confines the current flow through the pn junction to a thin stripe between the two end mirrors.

Because of this confinement, the current density in the active region is extremely high. This effectively lowers the threshold and the required operating current. A simpler and less expensive power supply can be used to run a stripe-geometry laser. With some such devices an applied forward current of less than 100 mA will be sufficient to generate a laser beam of several milliwatts.

The different types of injection lasers are compared in Fig. 12-6.

Because they can generate a low-power continuous beam, DH lasers are used in many low-power applications, including optical fiber communications systems, CD and video-disc reading, and laser printers.

EXPERIMENTING WITH INJECTION LASERS

Figure 12-7 shows a typical experimental setup to observe injection laser operation. Pulses of high current density are passed through the diode while it is immersed in liquid nitrogen. The light which emerges from the plane of the junction is usually in the near infrared region and can only be viewed indirectly through an infrared image converter (also called a snooperscope). Alternately, by passing the light through a spectrometer (an optical frequency analyzer), the spectrum of the emitter radiation can be determined. This spectrum shows a marked peaking coincident with the onset of laser action. To understand the operation of the injection laser, you should take note of the close analogy between the band structure diagram, familiar to those who work with semiconductors, and the energy-level diagram which represents the behavior of the atoms.

Of particular interest is the fact that combination of an electron-

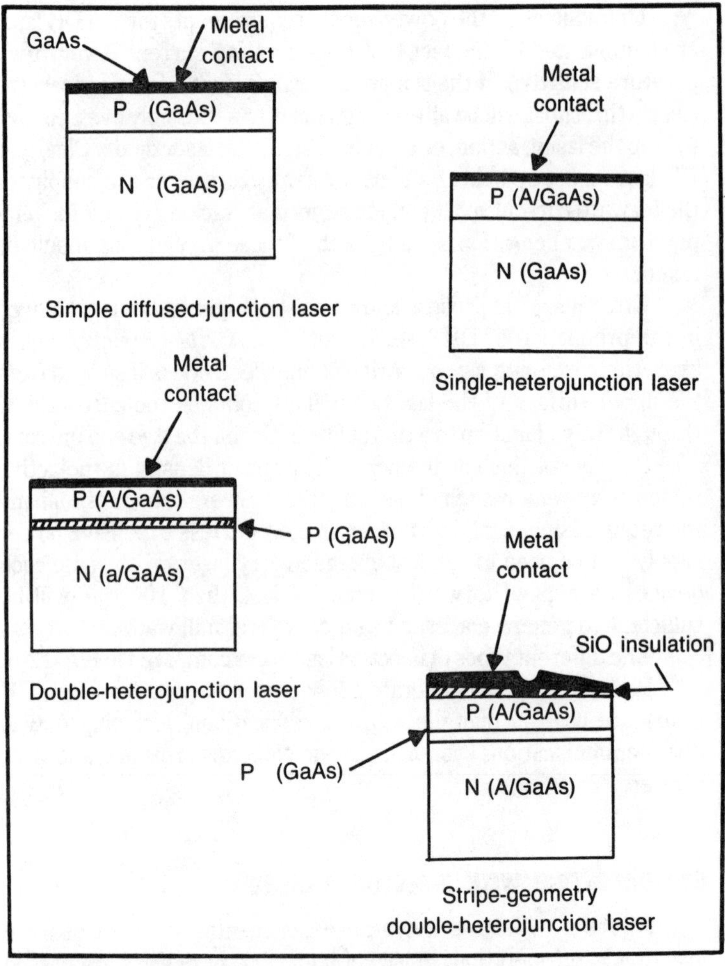

Fig. 12-6. Comparison of different types of injection lasers.

hole pair at a semiconductor junction is frequently accompanied by the emission of a photon. The photon's energy can be expressed as $E = hf = \Delta E_G$, where ΔE_G is the energy band gap. The energy gap is the difference in energy between the lowest level of the conduction band and the highest level in the valence band.

At normal temperatures, combination of carriers (conduction electrons and holes) can occur without any radiation. These combinations occur at sites where the periodic crystal structure is disturbed by the presence of impurities and defects in the crystal. Such sites

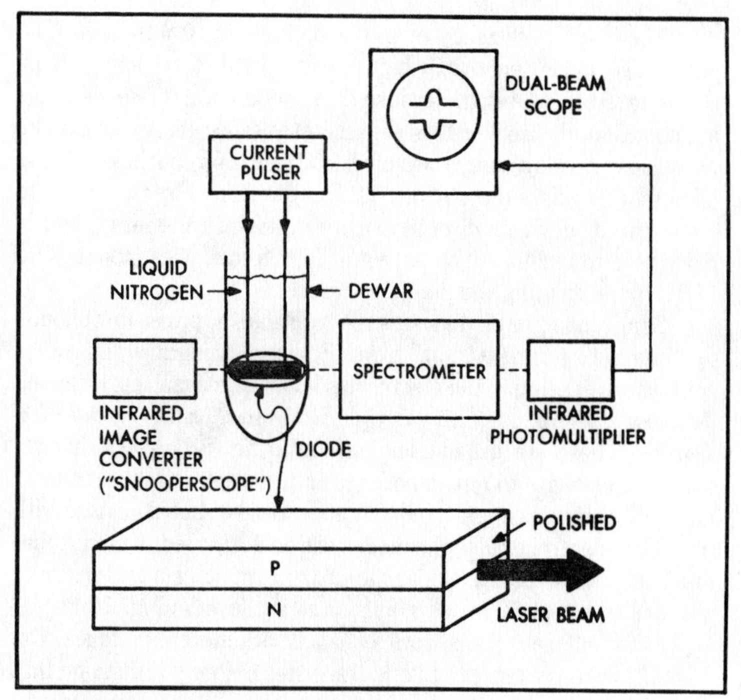

Fig. 12-7. Experimental setup for observing injection laser.

are called *traps*. The lifetime of an electron in the semiconduction band is strongly limited by the traps. For this reason, the conduction band is generally unsuitable as the metastable state, which is so crucial in obtaining population inversion.

When the temperature of the diode is lowered, the effectiveness of the traps is reduced, and lifetimes are considerably extended. In certain semiconductors at extremely low temperatures, the radiative process dominates combination. Under this condition, the situation is analogous to laser transition levels in the ruby crystal. In order to pump when this method is used, a region must be created in which conduction electrons and holes abound. The most elementary considerations show that this exists in the immediate neighborhood of the junction when a forward bias is applied. At extremely low temperatures and sufficiently strong forward bias, the region of the junction should be an amplifier for radiation satisfying $hf = E_G$. By cleaving and polishing the diode into a rectangle with faces perpendicular to the plane of the junction, a resonant cavity is formed, and oscillation can be observed.

The overall efficiency of the diode laser is 10% to 30%. The junction region is necessarily quite small, and the total power which can be radiated is limited. Because the conduction and valence bands are considerably broader than the discrete atomic energy levels, the emission contains a comparatively broad band of frequencies. These facts tend to offset the obvious advantages gained by the ability to pump the diode laser directly with wide electrical energy, and in being able to modulate the output at high frequencies (about 1000 MHz) by modulating the pumping current.

Semiconductor diode lasers are available, at prices low enough for home experimentation, from Edmund Scientific and other sources advertising in the electronics hobbyist magazines, including *Modern Electronics, Hands-On Electronics*, and *Radio-Electronics*. These are usually single-heterodyne (SH) lasers that require pulses of five to ten amperes to activate them. Only very short pulses (less than a couple hundred nanoseconds) should be used with the SH diodes, otherwise the diodes will be destroyed. A circuit that takes advantage of the avalanche characteristic of a transistor to produce pulses of 50 nanoseconds or so is shown in Fig. 12-8.

Since different transistors avalanche at different voltages, you might have to try more than one transistor before you find one that

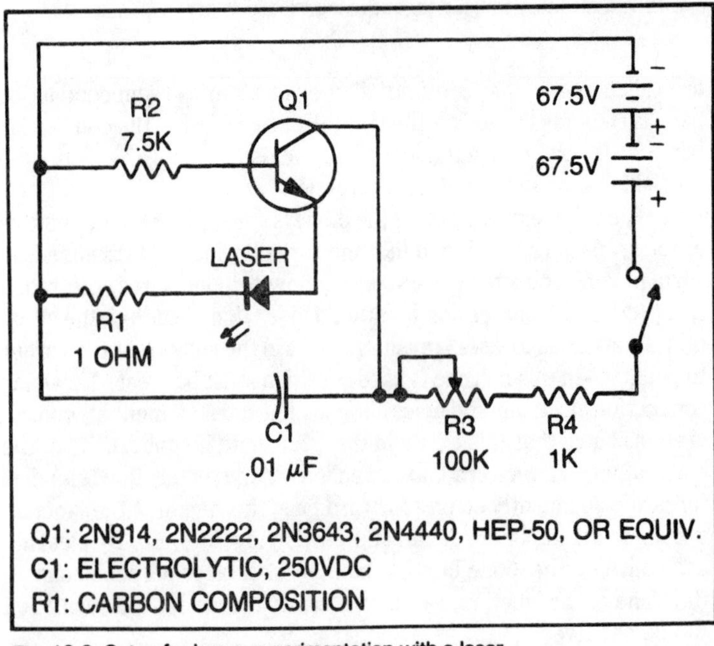

Fig. 12-8. Setup for home experimentation with a laser.

produces the current through the diode. In the circuit of Fig. 12-8, resistor R_1 is used to monitor the current that will flow through the diode. This should be a *carbon* resistor of one ohm. Since the resistance of R_1 is one ohm, the current through it is equal to the voltage across it. The voltage pulses across R_1, hence the current pulses through R_1, and can be measured with an oscilloscope of 15 MHz or better bandwidth. *Initially, during the transistor-selection process, replace the diode with a piece of copper wire*. Measure the pulses with an oscilloscope connected across R_1. Their height on the screen is the amplitude of the pulses in amperes. Before you insert the laser in the circuit, make sure that the transistor selected is one that produces pulses of somewhat less than three times the threshold current. Try an npn switching transistor such as the 2N914, 2N2222, 2N3643, 2N4400, or Motorola HEP S0011. For best results, keep the leads of the capacitor, transistor, laser, and one-ohm resistor as short as practicable.

Instead of using the elaborate equipment suggested in the preceding illustration, you can use a plastic viewing card. (You will not be able to see the beam without some kind of infrared viewer.) Kodak Special Product Sales (Rochester, NY 14650) has a plastic IR viewing card that sells for about $30.

In experimenting with a laser diode, you will note that the diode emits a rather divergent (broad) beam. You can narrow this down to a pencil of laser light with an *f*/1 lens (try Edmund Scientific for this). Before you begin your experiments, read the chapter on laser safety.

13

Laser Pumping

Pumping action of the different types of lasers is accomplished by several different methods, depending on the type of laser.

Operation of a gas laser (pumped by electron collision) requires a means of creating an electrical discharge in a tube containing low-pressure gas. This is done by either one of two methods, rf or dc pumping. In one case of rf pumping, an rf generator capable of supplying 50 to 100 watts of power is used. This supply is coupled to an electrode, or electrodes, spaced around the outside of the gas-filled tube. Coupling is provided by a matching coil specially designed for maximum power transfer to the electrodes (see Fig. 13-1). For an approximately one-milliwatt output laser (typical helium-neon operation), pumping power levels run from 20 to 50 watts at about 30 MHz.

Alternately, the discharge can be maintained by supplying a sufficiently high dc field between electrodes contained in the laser tube. Heating the cathode with a filament supplies electrons for the collision process. The necessary potential depends on the separation between the electrodes and the tube diameter. A potential drop of four kilovolts is typical in a laser tube four millimeters in diameter and one meter long.

Pumping a solid laser rod, such as ruby, is accomplished optically. To obtain sufficient light power for inversion, it is usually necessary to fire a high-power xenon-filled flash tube. The flash tube is optically coupled by mirror geometry to the laser rod. Operation of

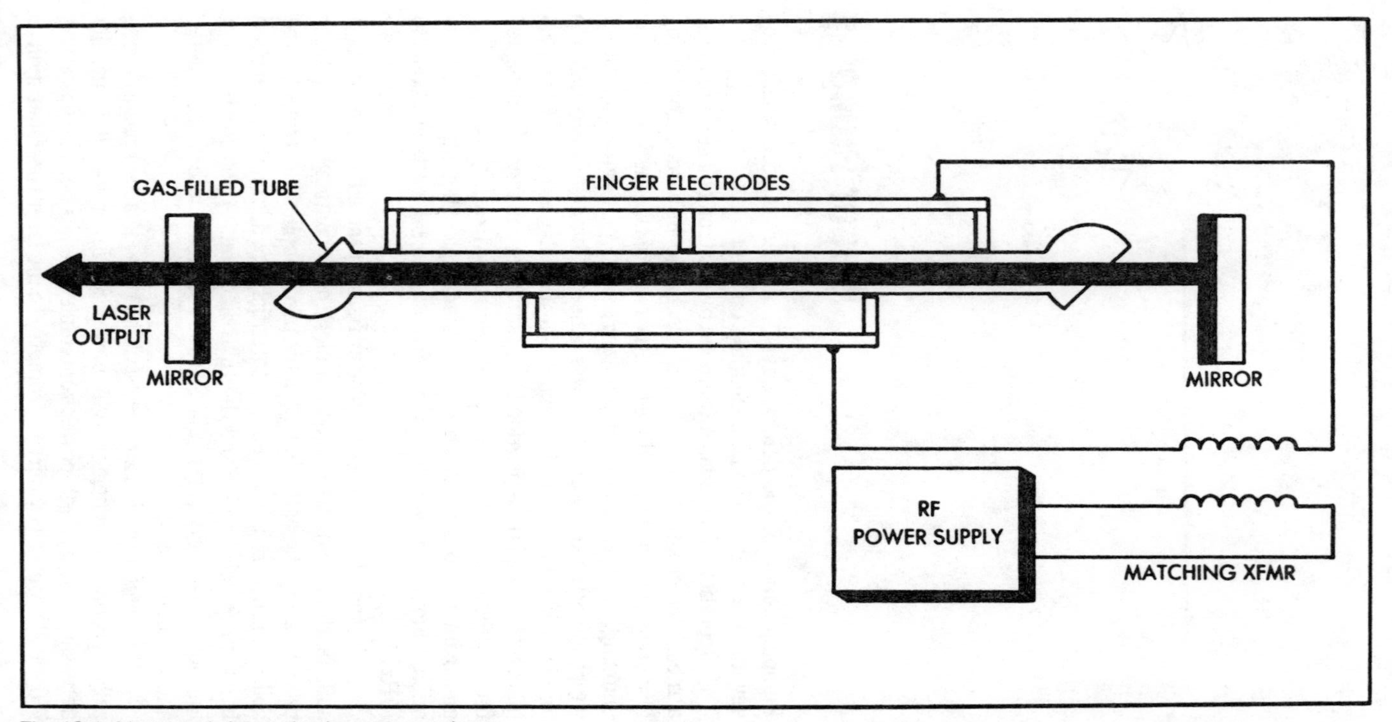

Fig. 13-1. Method of rf pumping in a gaseous laser.

the flash tube is similar to that of a thyratron. A high-voltage trigger breaks down the xenon gas between two electrodes, which have been maintained at different potentials. Upon breakdown, the lamp resistance drops to a low value of one ohm, and high peak currents flow.

The magnitude and duration of these currents depend upon the capacity and voltage of the supply; that is, the peak current is $I_P = V/R$ and the duration, T, is equal to RC, where V is the supply voltage, C is the capacitance, and R is the resistance of the flash tube. Flash tubes are generally rated in maximum watt-seconds (or joules) per current pulse. Typically, these values are quite high. For example, 250 watt-seconds are necessary for a small ruby laser, and up to 12,000 watt-seconds may be needed for a large ruby laser. Consider-

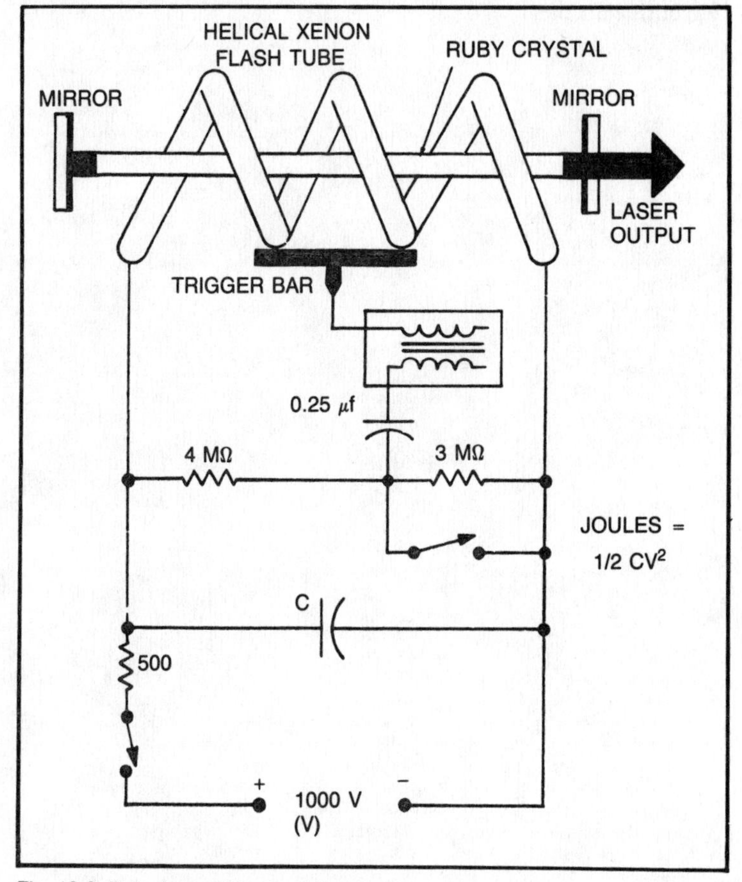

Fig. 13-2. Ruby laser with xenon flash tube.

able energy must therefore be stored in large capacitor banks before discharge. The large values of stored energy at high voltages make safety an important consideration in handling the power supplies for even small ruby lasers. Figure 13-2 shows a typical power supply schematic for small ruby laser.

Operation of the injection laser on a pulsed basis (typical usage) is similar in many respects to firing a flash lamp. High peak currents of up to 70 amperes are necessary to create inversion, and the forward resistance of the diode is considerably less than one ohm. There is no provision for triggering the current. A transistor or a silicon-controlled rectifier (SCR) must be placed in series with the diode. Direct modulation of forward current, and hence laser output, can be applied by coupling via a T-coupling from an rf generator to the pumping supply. Modulation rates of up to 1000 MHz are possible in this manner.

Laser Communication

Since the methods used for modulation and demodulation are different than those used for conventional electronic systems, a brief description is provided in the following paragraphs.

MODULATION

The laser, by virtue of its coherence, offers the possibility of using carrier frequencies of 10^{14} to 10^{15} hertz for communications.

Theoretically it is possible to modulate a 10-MHz carrier with approximately two-hundred 20-kHz audio channels without overlapping. A laser channel at a center frequency of 10^{15} hertz will permit ten million times that number. A single laser is therefore capable of simultaneously emitting 30 billion 20-kHz audio channels. To develop the laser to its full potential as a communications carrier, means of coherent modulation and demodulation which are compatible with the tremendous allowable bandwidth must be provided.

Although the energy levels appearing on the energy level diagram for an atom are quite well defined and narrow, the stimulated emission is not strictly monochromatic. The spreading of the spectrum of radiation is due to several causes, the most significant of which is simple Doppler shifting. This phenomenon is essentially the same as that encountered in sound waves. The motion of the source relative to the receiver causes a shift in the received frequency proportional to the relative velocity. The atoms in a gaseous discharge have random velocities because of their thermal motions, which can be as high as 10^6 centimeters per second. These random velocities shift the emission and absorption frequencies of the atom over a 100 MHz frequency band.

Figure 14-1 shows the typical spectrum of a helium-neon laser. Note that although the individual oscillator resonances are narrow, the laser oscillator is operating at several frequencies simultaneously. To achieve a single oscillator frequency, the output must be passed through a narrow-bandpass optical filter. This is done by coupling to another high-Q cavity of a different length, which is resonant for one of the oscillator outputs. Since the lengths are different, the spacing between resonant modes is different, and the second cavity eliminates the other modes. The laser output, after passing through the narrowband filter, contains only a single frequency.

Applying this laser output to communications requires a means of wide-band modulation and demodulation. Direct modulation of the pumping source (flash lamp or gas discharge) in gaseous or solid lasers, which is limited by thermal inertia to a few kilohertz, is obviously insufficient. Other techniques to be considered can be divided into two groups. These are internal modulations in which the modulation is impressed upon an element of the feedback loop, and external modulations in which the modulation is impressed directly upon the oscillator (light-beam) output.

Various potential wide-band modulation schemes have been proposed, several of which have been successful over several gigahertz. The following paragraphs consider only amplitude modulation by means of varying the oscillator cavity length.

To obtain a fast shutter which can be directly operated by a voltage, a Kerr cell can be used. A *Kerr cell* consists of a special clear

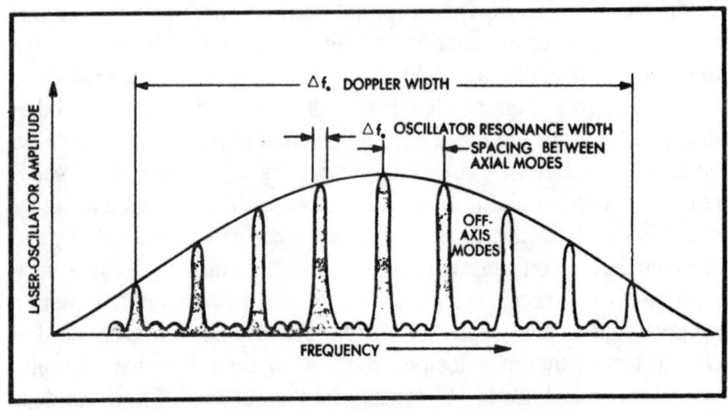

Fig. 14-1. Oscillating spectrum of oscillation helium-neon laser.

liquid (for example, nitrobenzene) between two electrodes. When voltage is applied to the electrodes, the liquid acts to rotate the plane of polarization of light passing through it. The amount of rotation is proportional to the length of the cell and the square of the applied voltage. By modulating the voltage across the two electrodes, a linearly polarized beam will become polarization modulated. To convert polarization modulation to amplitude modulation, the light is passed through a polarizer. The transmission, T, of a linear polarizer for linearly polarized light is simply $T = To \cos^2\theta$, where θ is the angle between directions of polarization, and To is the transmission when they are perfectly aligned.

Figure 14-2 shows the operation of a Kerr cell with a laser transmitter. Modulation by this means is limited to a few megahertz by the time constant of the Kerr cell itself. To achieve higher modulation frequencies, the interaction distance must be increased by forming a transmission line embedded in a time-varying dielectric, where the Kerr cell liquid acts as the dielectric.

Frequency modulation of the laser output can be accomplished by changing the cavity length, thereby varying the frequency of the resonant modes. For a helium-neon laser, a change of cavity length of one-half wavelength (to the next resonant mode) corresponds to a frequency shift of 150 MHz. A half wavelength for a helium-neon laser is only 3×10^5 centimeter. The change of length necessary for wideband modulation is, therefore, quite small. Such small displacements can be conveniently provided by a piezoelectric material.

A piezoelectric material, such as quartz, is a material which contracts and expands along a dimension perpendicular to the direction of an applied voltage. If the applied voltage is a varying voltage, and one face of a thin slab of piezoelectric material is secured firmly to a stationary object, the exposed surface will undergo a vibration. This vibration can be transferred to a variation in cavity length by mounting one of the cavity mirrors on this surface. The motion of the mirror modulates the cavity resonance and thus frequency modulates the laser output. This method of modulation is suitable up to modulation rates of 1 MHz, at which point the inertia of the mirror becomes the limiting factor. It is not practicable to consider the extension of such a method to a very high frequency, since the range of modulation is limited by the Doppler width of emission line.

To detect coherently a modulated carrier, the received signal must be compared to a local oscillator signal in a nonlinear element (mixer). In the case of a laser communications link, the local oscillator is provided by a second laser at the receiver, operating at the same

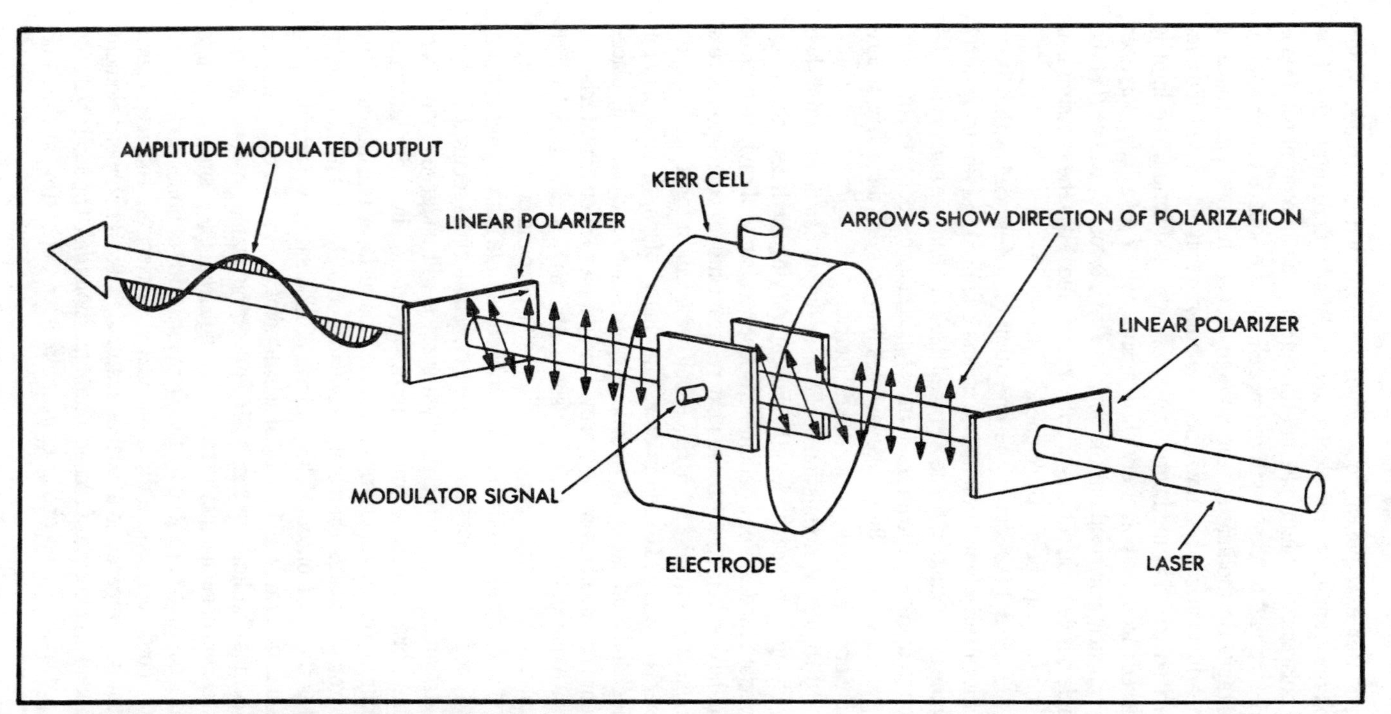

Fig. 14-2. Amplitude modulation of laser beam with Kerr cell.

wavelength as the transmitter. The nonlinear element can be any material with nonlinear response at the carrier wavelength. It can be the surface of an ordinary photodetector. All photodetectors are essentially square-law devices; that is, the voltage developed is proportional to the intensity of light at the sensitive surface, which, in turn, is proportional to the square of the electric field.

If the two inputs are mixed at the sensitive surface, the resultant response contains a term whose amplitude is proportional to the product of the local oscillator signal and incoming signal strength. The frequency is equal to the difference in frequency between the output of the local oscillator and the signal. By frequency locking the local oscillator to the transmitter, the carrier is suppressed and the information (AM or FM) appears directly as voltage at the terminals of the detector.

Note that the signal bandwidths concerned must exceed 1000 MHz for effective use of the laser carrier frequency. This places a requirement on the frequency response of the photodetector which can be met only by means of specialized construction of the detector element.

MODULATION OF A GaAs INJECTION LASER

Gallium arsenide (GaAs) has a minority carrier combination lifetime on the order of 10^{-9} second. Since the radiative combination lifetime decreases in the stimulated emission mode, it is theoretically possible to modulate gallium-arsenide lasers at the microwave frequencies.

A block diagram of the modulation and demodulation arrangement is shown in Fig. 14-3. A 0.1-microsecond, 12.5-ampere video pulse is used to drive the gallium-arsenide diode into the laser action mode. A one-milliwatt CW signal from the microwave generator is gated by the microwave modulator to produce 0.1-microsecond pulses synchronized with the video pulses. The microwave pulses are amplified in the traveling-wave tube to a peak level of one watt, and added to the video pulses in the dc monitor T-coupling. The sum is fed to the laser-mounting termination.

The termination is shown in Fig. 14-4. A 50-ohm carbon composition resistor is used as a series-resistive match. The laser diode is pressed between the resistor end cap and the copper cooling block, which is immersed in liquid nitrogen. The spring loading maintains constant contact pressure under thermal contraction of the diode. The laser light beam is emitted through the hole in the outer conductor.

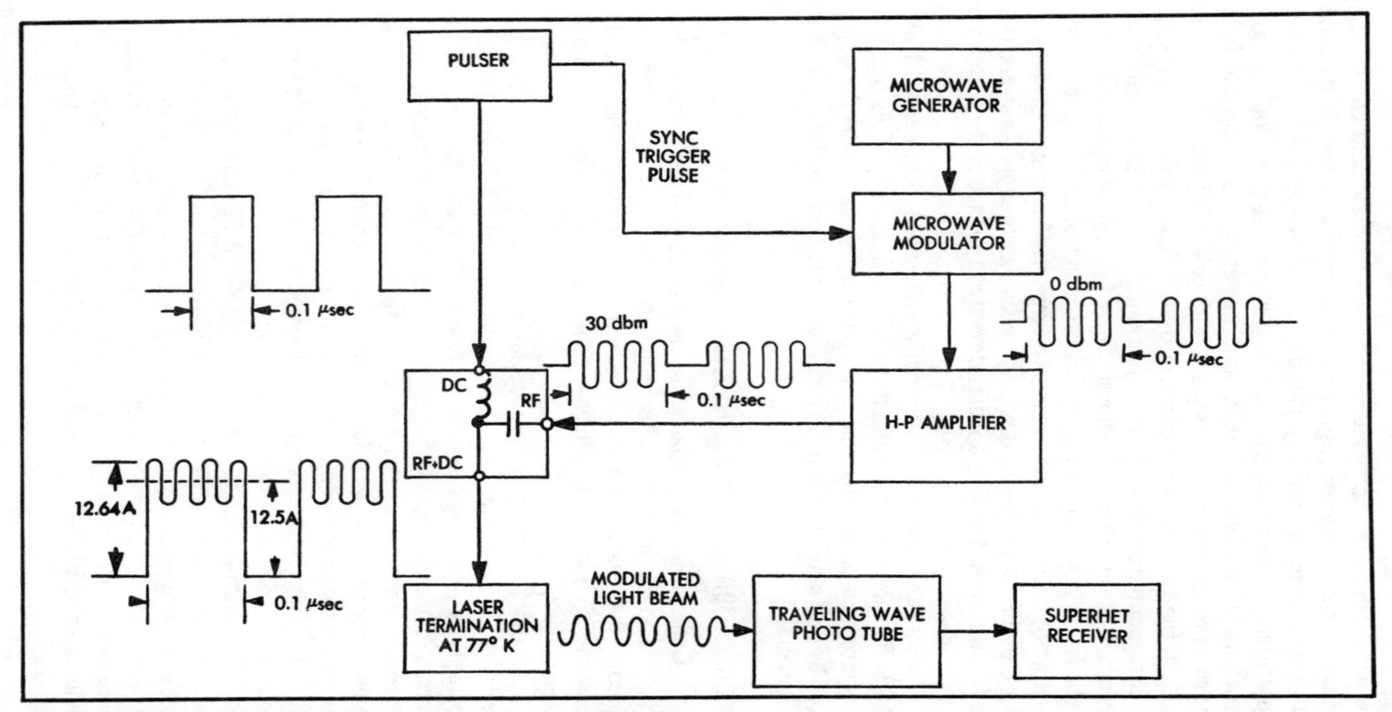

Fig. 14-3. Modulation and demodulation arrangement for an injection laser.

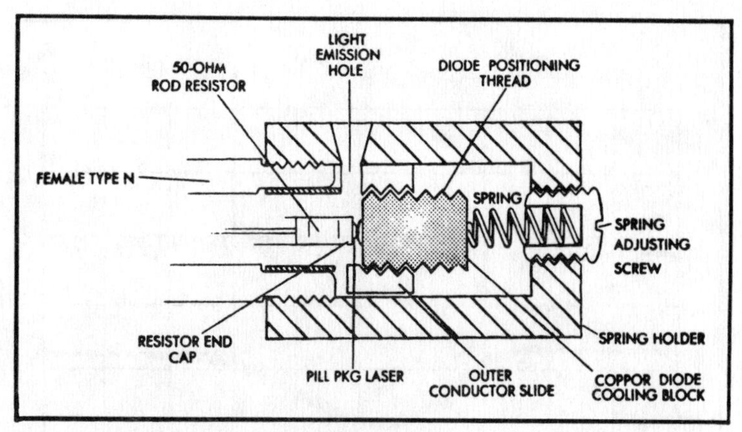

Fig. 14-4. Termination for microwave modulation.

The laser is constructed in the low-inductance pill configuration shown in Fig. 14-5. The crystal is made by diffusing zinc into tellurium-doped gallium arsenide with a net impurity concentration of 3×10^{-17} per cubic centimeter. The junction area is 2×10^{-3} per square centimeter, and the laser action threshold for this particular laser is 10.5 amperes (5.25×10^{3} amperes per square centimeter).

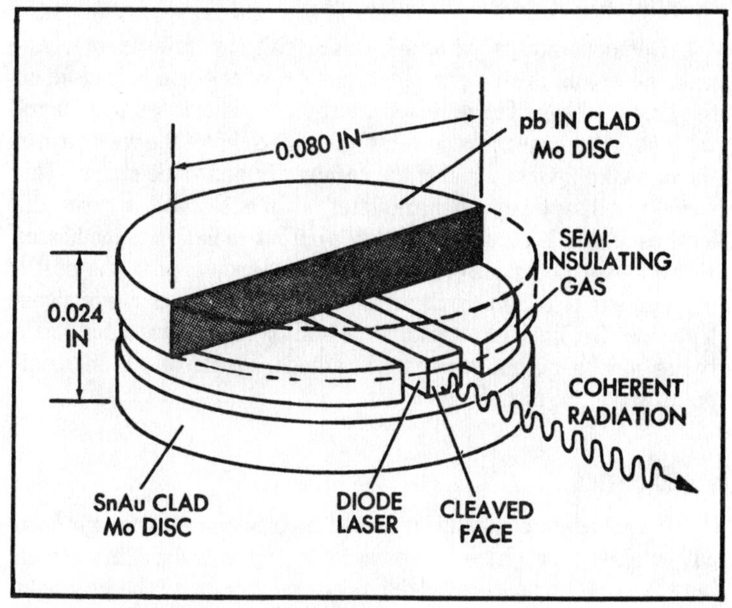

Fig. 14-5. Pill laser-diode configuration.

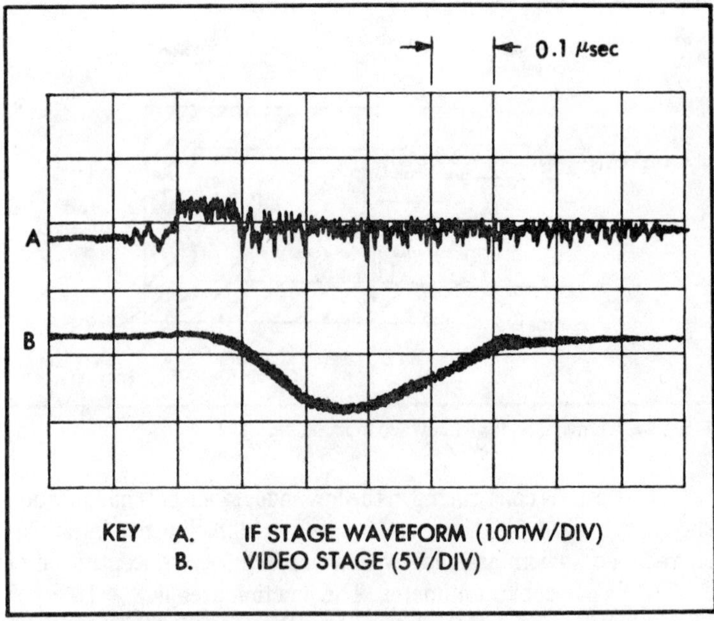

Fig. 14-6. Demodulated waveform.

DEMODULATION

Demodulation can be accomplished with a microwave traveling-wave phototube with a photosurface. The laser beam is focused on the photocathode. The detected microwave signal is fed to a microwave superheterodyne receiver. Figure 14-6 shows the waveform of the detected 2-GHz signal at the receiver IF and video stages. This waveform disappears when the laser beam is blocked or when the local oscillator is detuned. The laser output radiation is modulated with a modulation index of about 7% and produces a peak photocathode current of 0.5 microamperes. The detected microwave signal level is −73 dBm. The phototube calibration data indicate that this is the obtainable output power for a cathode current of 0.5 microamperes and a 7% modulation index.

FIBER-OPTICS

Laser-beam communication can be transmitted through open air, but only over a line-of-sight path. Any opaque object in the path between the transmitter and the receiver will block off the laser light beam.

One popular solution is to use an optical fiber cable to conduct the laser beam communication signals over any predetermined path.

A fiber-optic cable functions rather like an ordinary copper cable, except light waves are conducted rather than electrons.

An optical-fiber cable is made up of a number of transparent glass or plastic fibers, some no thicker than a human hair.

Ordinarily light waves (including laser light) travel only in straight-line paths. The light waves can be forced to follow any curve, or even sharp angle in a optical fiber cable, thanks to refraction. The refractive index of the fiber's core is slightly higher than that of its cladding. This effectively confines the light to the interior of the fiber, regardless of any bends or curves. This is illustrated in Fig. 14-7.

The transmitter is a modulated light source, often a laser, at one end of the optical fiber cable. Some sort of photosensitive detector serves as the receiver at the far end of the cable.

An ordinary light source can be used in this type of application, but a laser offers some significant advantages. The intense power and coherence of laser light gives increased range, and more data can be simultaneously transmitted over a single cable.

Optical fiber cables can be used in almost any application where an electrical conductor cable would normally be used. Fiber-optics has many advantages over standard electrical conductors.

For one thing, optical-fiber cables tend to be cheaper than coaxial cables, and there's no reason to assume that the price difference won't be even greater in the future. Coaxial cables are normally made of copper wires and the dielectric is made from petroleum-de-

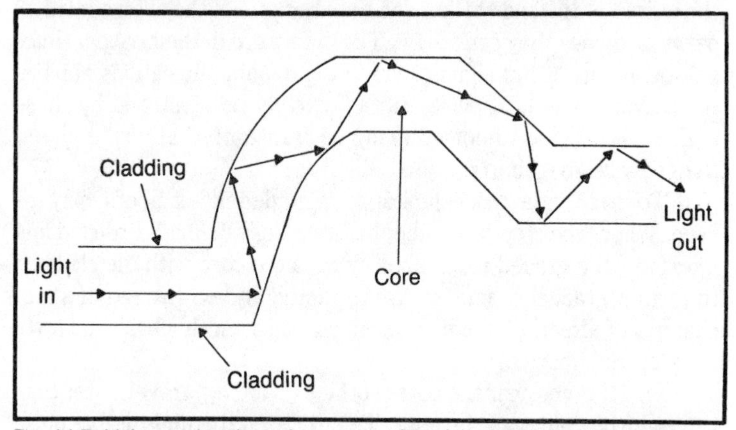

Fig. 14-7. Light passing through an optical fiber follows any bends or curves of the fiber because of the different refraction indexes of the core and the cladding.

rived plastic. The raw materials are inherently expensive. The basic raw material for glass optical fibers is simple sand, an ingredient which is likely to remain plentiful and cheap.

A single glass fiber the thickness of a human hair can carry more information than 900 pairs of copper wires, which would require a cable at least as thick as the average person's wrist. In addition, that hair-thin strand of glass fiber has a higher tensile strength than a steel wire of the same thickness.

Perhaps the most significant advantages of fiber optics stem from the fact that the conducting cables are insulators rather than conductors. There is no shock hazard or risk of short-circuits with optical fiber cables. They will not generate any sparks. Also, optical fibers are not susceptible to electromagnetic interference, and they do not attract lightning strikes. It is inherently difficult to jam or intercept communications in a fiber-optics system.

Typically a modulated laser light beam can travel more than ten kilometers through an optical fiber cable without needing any external amplification stages. If longer transmission distances are required, strategically spaced repeater stages can be used to boost the signal level as needed.

The most basic type of optical fiber is the step-index fiber. This is the type illustrated in Fig. 14-7. There is a sharply defined transition between the core and the cladding. The refractive index abruptly changes, and the light beam is reflected at a definite angle.

This type of optical fiber is popular because it is easily manufactured, but it is not without its disadvantages. Two or more light waves simultaneously fed into one end of the fiber might not reach the output at the end at the same time because of the different travel paths or modes they can follow. The difference in their output times will be slight, but of significance in many communications applications. Narrow optical pulses can effectively be stretched by these path delays. Data cannot normally be transmitted at a rate higher than several megahertz.

To overcome these limitations, graded-index fibers may be used. Where the step-index fiber has a strongly defined core/cladding interface, the graded-index fiber merges the core with the cladding to form a gradual change in the refractive index. Instead of sharp changes of direction, the light beam curves smoothly back and forth through the fiber core.

Light waves near the outer surface (cladding) travel faster than through the center of the core. This reduces the undesirable pulse stretching effect and permits data rates up to several-hundred megahertz.

The basic type of optical fiber is called multimodal. Several light rays are transmitted through the fiber at any given time. The various rays have slightly differing frequencies. It is hard to avoid at least some interaction between these varying light rays.

A fairly recent development is the monomode fiber which is designed to transmit just a single light ray configuration.

To take advantage of the monomode fiber's special advantages, an extremely pure laser must be used as the light source. In theory, all lasers emit a pure light beam at a single frequency. But in practice, standard lasers emit a blurred spread of frequencies.

The frequency of the emitted light from a laser depends on the lasing material used. Any substance has a specific "signature" frequency. However, a completely stable, single frequency can only be achieved if all the atoms in the lasing material are perfectly still. Except at a temperature of absolute zero, this never happens. The atoms will be moving about, blurring their "signature" frequency because of small Doppler shifts. Rather than emitting a true single frequency, the laser is slightly impure, emitting a band of frequencies.

Bell Laboratories has developed an extremely small laser that generates single frequency light beams with very high purity. This new laser device is also tunable, so any of several discrete frequencies may be used, any one of them with a high level of purity. The device is called the *cleaved coupled-cavity laser*. It is sometimes shortened to the C^3 laser.

A mirrored cavity amplifies only specific frequencies (actually a small range of frequencies). Which frequency will be amplified is determined largely by the length of the cavity. The C^3 laser employs two separate cavities of slightly differing length. They are coupled optically so that only light at a frequency that can be passed by both of the cavities individually will get through the pair. Each cavity therefore limits the other's permissible range of passed frequencies.

To form a C^3 laser, a semiconductor laser is precision cut into two sections which are then rejoined along the cleaved surface. The two sections are now electronically isolated from one another, but they are optically coupled.

Varying the electrical current applied to the lasing material changes the optical properties, and thus the emitted frequency of the laser. This is how the C^3 laser is made tunable.

Since the optical fibers are most transparent to light frequencies of about 1.5 microns (in the infrared region), this is the frequency the C^3 laser is generally tuned for.

The combination of the cleaved coupled-cavity laser and monomode fibers increases the unboosted (no repeaters) transmission range for data communications. The record is now 119 kilometers. In addition, the error rate is significantly reduced, almost to the point where it can be ignored. If an entire encyclopedia is transmitted through this type of system, the anticipated error would amount to no more than a single character being erroneously capitalized.

Since the tuned frequency of the C^3 laser is current controlled, this device can switch between frequencies at an extremely high rate. Ten discrete frequencies can be switched a billion times per second. By switching between various frequencies in a specified pattern, multiple messages can be encoded on a single beam of laser light.

Laser Radar

Although still in the early stages of development, optical radar already provides a coherent light beam 100 times narrower than beams of ordinary microwave radar equipment. Radar pictures made with laser radar are exceptionally sharp, revealing far more detail than previously obtainable. For example, an optical radar demonstrated recently can distinguish between two adjacent ten-foot objects at five miles. Enhanced resolution is obtained with equipment of small size and light weight, because light waves can readily be focused by lens-and-mirror arrangements. Compare this with ordinary radar microwaves that require immense antennas to focus the radio energy into comparably narrow beams.

Laser radars should be especially valuable for use in space, where size and weight are critical, but where there is no absorption or scattering of light rays by rain, clouds, or by the atmosphere itself. It follows that in many space applications, stray sunlight, which might overwhelm the light pulse detected by the laser radar receiver, can be avoided. With the exceptionally narrow light beam practical with laser methods now, it is feasible to illuminate a spot only two miles in diameter on the moon's surface for accurate mappings of lunar topography.

PRINCIPLES OF OPERATION

For radar operation, a short burst of red light from the laser is aimed or directed at a target to be detected. Light reflected back from the target is gathered by a telescope and detected by a light-sensitive phototube. Distance to the target is determined by measuring the time required for the light burst to make the round trip. The elements of this method are shown in the block diagram of Figure 15-1. The electrical signal resulting from the light pulse returned to the receiver is displayed on a cathode-ray oscilloscope. Time delay for distance measurement is obtained by comparing the received signal with a sample of the transmitted light burst as monitored by a reference phototube. Figure 15-2 illustrates a laser receiver.

For ideal microwave radar performance, the transmitter output should be composed of a single short pulse. The pulse preferably should last for a few microseconds, or a few millionths of a second.

One of the principal components of the laser ranging equipment is the receiver. Here light is gathered by another telescope, with a five-inch lens. The larger the diameter of the telescope, the more sensitive the receiver. Light entering the telescope is first passed through a red filter which does not impede the 6943-angstrom radiation from the laser, but which does reject 99.9° of the light from all other ordinary sources, such as sunlight, light from incandescent lamps, etc. Light passing through the filter is then directed at a very sensitive phototube, where it is converted into an electrical signal for display on the oscilloscope. A photomultiplier type of phototube provides a very sensitive detector.

Rough calculations indicate that only two or three orders of magnitude improvement in the system would be required to obtain

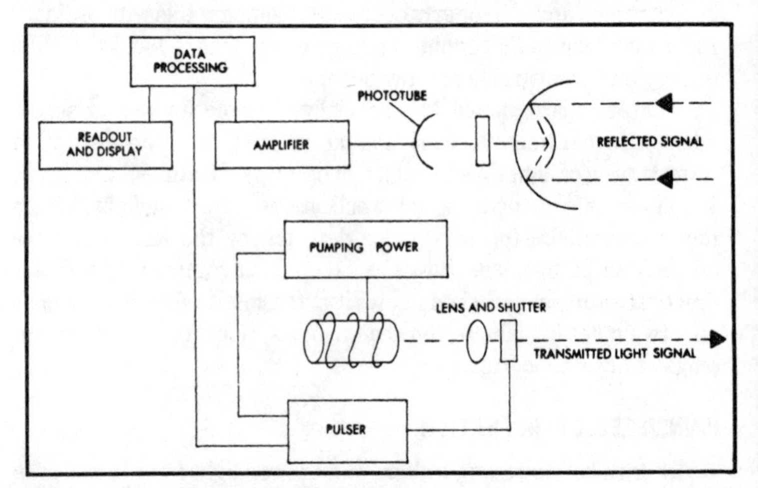

Fig. 15-1. Block diagram of a laser radar set.

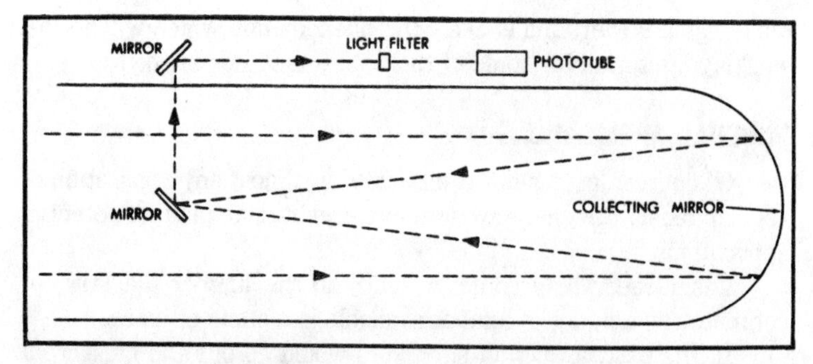

Fig. 15-2. Laser radar receiver.

detectable pulses from the surface of the moon. Refinements already well under way include high-power lasers operating with short bursts (to reduce problems from back scatter from dust and water particles in the air), more sensitive receivers, and improved optical systems.

NOISE EFFECTS

As with any radar system, the basic limitation in performance that can be achieved by the laser radar is determined by the noise level through which the signal must be detected. Three types of noise affect the system. These are internal or receiver noise, external noise caused by unwanted light entering the receiver through the red filter, and fluctuation or shot noise in the signal itself.

Receiver noise is caused by dark current (current flowing in the absence of any light) in the phototube. This current is produced by the thermionic emission from the sensitive photosurface and can be essentially eliminated through refrigeration of the photocell. Dark current noise at room temperature is low enough to be neglected with present-day systems.

Secondly, for daytime operation on the earth's surface, that part of the scattered sunlight which can pass through the red filter imposes the limiting noise for the equipment.

The third type of noise comes from signal fluctuation—a new kind of noise not encountered in ordinary radio—which would still be an ultimate limitation for a laser radar even if the other two sources were completely eliminated. Fundamental in nature, it is caused by the random rate of arrival of the light photons (basic unit of light energy) which make up the received signal. Noise of this kind can be understood in a qualitative way by recognizing that at least one photon must be detected at the receiver during the time interval when a signal is expected. It represents an ultimate minimum detectable signal power, and is likely the phenomenon which will be the working limitation for some of the optical methods of the future.

UNUSUAL APPLICATIONS

Of course, laser radar can be used in almost any application of regular radar. But there are also some additional, unique, potential applications.

Laser radar may come in handy in the lumber industry, to estimate the amount of lumber available in a stand of trees.

In the past, a team of surveyors would go out and measure a sample of the trees (about 5% to 10%). This is time consuming. For a stand of just a few hundred acres, the inventory can take two or three days.

The results of the sampling are then used to estimate the total amount of lumber available in the area. Even if a computer is used, this still takes time, and adds up to a major expense. The profit-margin in forestry is very low, so there is high incentive to try to avoid such expenses.

A new technique using a laser is currently under development, and looks quite promising.

A small plane is flown about 600 feet over the stand of trees. A low-power laser aboard the plane projects a beam that hits the tree-tops, is reflected from the ground, and bounces back to the plane. In other words, the basic radar method is used.

The laser's power is low enough that it does no harm to the trees, or anything else that might be in the vicinity. There appear to be no ecological problems with using this method.

The plane also carries a microcomputer which keeps track of how long it takes the laser beam to return.

After the flight, this data is fed into a larger computer on the ground. A line drawing of the trees is generated from the data obtained during the flight. The drawing indicates the height and crown diameter of the trees. A reasonably accurate estimate of the lumber available in a tree can be made from just these two parameters.

The system is still under development. It isn't quite accurate enough for industrial use yet, but the problem appears to be solvable, and this measurement technique should become a reality in the near future.

16

Laser Gyroscope

Counterrotating coherent light beams using a newly developed technique can provide sensing of rotation rate with respect to an inertial frame of reference. The light beams are produced in a traveling-wave laser that forms a closed circuit. Called a ring laser, the device holds promise for inertial guidance methods that would be more stable, have less drift, be simpler, cheaper to produce, and be more sensitive than present methods using gyros.

If the ring laser were to be mounted in a space vehicle that is not rotating relative to the stars, the photodetector would see no difference in frequency between the two light beams. Even the tiniest rotation of the vehicle would cause one beam to travel slightly farther than the other around the ring. The frequency of the two beams would then vary slightly, and the difference would be proportional to the rate of rotation. Measurement of the rate could then be used to direct the attitude of the vehicle to maintain its correct flight path. The angular displacement of the vehicle, or how far it has moved from its correct flight path, would also be measured by the ring laser.

Development of the ring laser has roots in optical experiments conducted half a century ago. In 1914, the French physicist, G. Sagnac, used a ring configuration approximately one meter square to form an interferometer (see Fig. 16-1). His device had poor sensitivity, and was barely able to detect a rotation rate even as fast as 43,000° per minute. His interferometer formed fringes, and he dis-

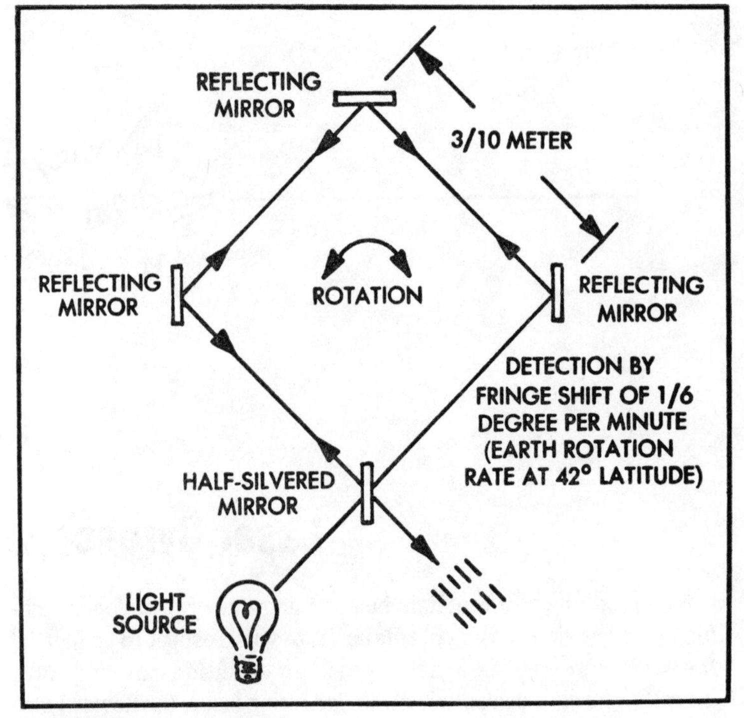

Fig. 16-1. Sagnac experiment.

covered that the fringes shifted sideways slightly when the loop was rotated.

In 1925, two American experimenters set up a loop that had better sensitivity, but was physically huge (see Fig. 16-2). The loop, employing light-tight, evacuated pipes, measured 0.4 mile by 0.2 mile. The two experimenters were able to detect the rotation of the earth, approximately one-sixth of a degree per minute by observing the fringe shifts, which were proportional to the rotation of the earth. The loop was stationary on the surface of the earth and could not be speeded up or slowed down. To obtain an independent reference of fringe shift, the physicists set up a smaller loop inside the larger loop, and compared the magnitude of the two fringe shifts with the areas of the two loops to show that the constant proportionality was caused by the earth's rotation.

The ring laser consists of four helium-neon gas tubes, each about three feet long. These are positioned to form a square, with Brewster angle windows and four external corner mirrors. The square ring composes the cavity of the laser so that light beams

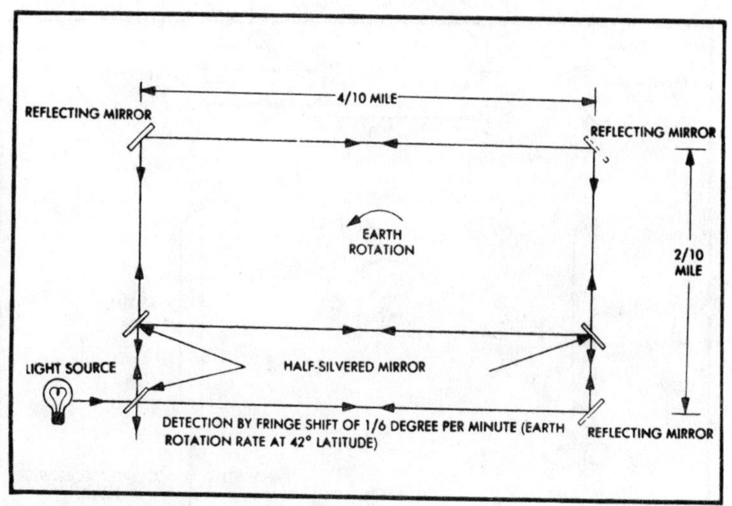

Fig. 16-2. Optical-loop experiment.

traveling around the circumference oscillate independently of each other. Their frequencies are determined by the length of the paths they travel.

When the ring (or its mounting) is rotated slightly, one light beam travels a slightly greater distance than the actual circumference to close on itself; the frequency of this beam is lowered. The opposite beam must travel a slightly shorter distance, so its frequency is raised. The two beams are mixed by optical heterodyning in a silver-cesium-oxide photodetector. The resultant beat frequency is proportional to the rotation rate of the ring.

The output of the photodetector can be integrated by circuitry that clips the sine waves, differentiates them, and then counts cycles. Thus, rate output is a measure of angular displacement, accurate to within 0.015°.

The equipment illustrated in Fig. 16-3 uses a meter-square ring laser oscillating at the helium-neon 1.153-micron wavelength. Excitation of the gas is accomplished by 27 MHz rf energy. Beat frequencies ranging from 500 Hz to better than 150 kHz were observed in the first experiments, with corresponding rotation rates of two degrees per minute to 600 degrees per minute. These are not the physical limits of the sensing effect, but are the limits set by equipment and techniques in the first demonstration setup.

The ultimate limits of sensitivity have not yet been determined, but developers are working to bring the 500-Hz sensitivity down as

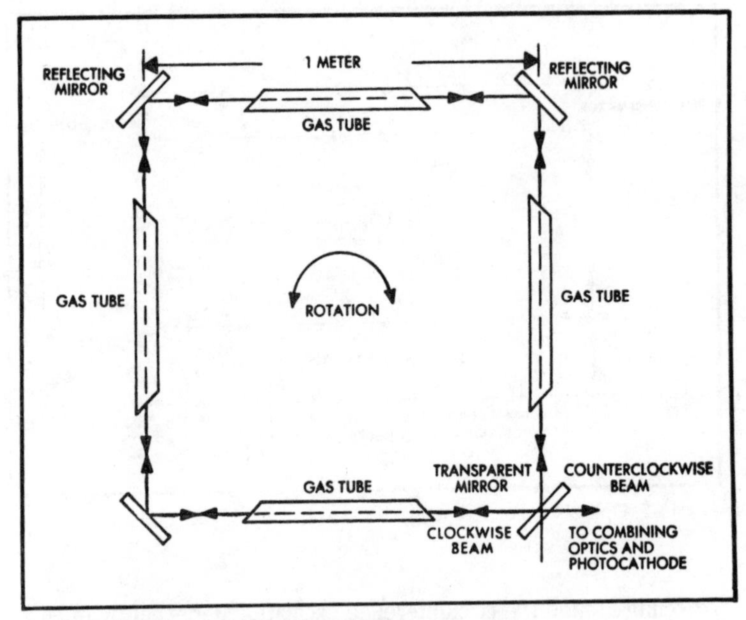

Fig. 16-3. Ring laser rotation-rate sensor.

close to zero drift as possible. The range of sensitivity is adjustable by varying the size and form of the ring.

The beat frequency for a square ring is $\Delta f = \omega L/\lambda$, where ω is the angular velocity of rotation, L is the ring perimeter, and λ is the wavelength. Maximum sensitivity is obtained when the axis of rotation of the sensor is perpendicular to the plane of rotation of the ring.

Present self-contained rotation sensors are gyroscopic and depend on conservation of the angular momentum of a rotating mass. This mass is subject to acceleration and gravitational forces. The laser rotation-rate sensor is free of these mass effects. It is a motionless strapped-down unit that requires neither bearings nor other mechanical moving parts. The simple construction of the sensor promises a low unit cost on a production basis.

Since the output of the ring laser sensing unit is basically digital, rather than analog like the output of a gyroscope, instrumentation is simple. The beat frequency can easily be clipped, differentiated, and counted. Instantaneous readout of both angular velocity in hertz and angular displacement in total cycles is available. The output is much more adaptable to instantaneous readout than the angular-shaft rotation output of a gyroscope.

17

Lasers in Space

This chapter describes two promising applications of lasers involving satellites: geodynamics and communications.

SATELLITE LASER RANGING

In satellite laser ranging, the distance to a satellite is determined by measuring the round-trip travel time of a pulse of light between a laser source and a satellite, which reflects it. In the tracking systems used by NASA, the pulses and measurements occur once per second. If two such systems track a satellite simultaneously from different sites (Fig. 17-1) it is possible to calculate the relative locations of the two systems, provided that the motion of the satellite is known. If the systems are on opposite sides of an earthquake fault, as in Fig. 17-1, it is possible to measure the slippage of the plates (large land masses) on either side of the fault.

Other scientific benefits to be gained from satellite laser ranging include more data on polar motion, solid earth tides, and variations in the earth's rotation. Scientists are interested in these matters because they will yield a better understanding of geodynamics, but as often happens, the scientific research can lead to commercial applications that are important to everyone.

LASER-RANGING STATION

In a mobile laser-ranging station, a ruby laser is used. This ruby laser has a power of 0.25 joule, a repetition rate of one pulse per

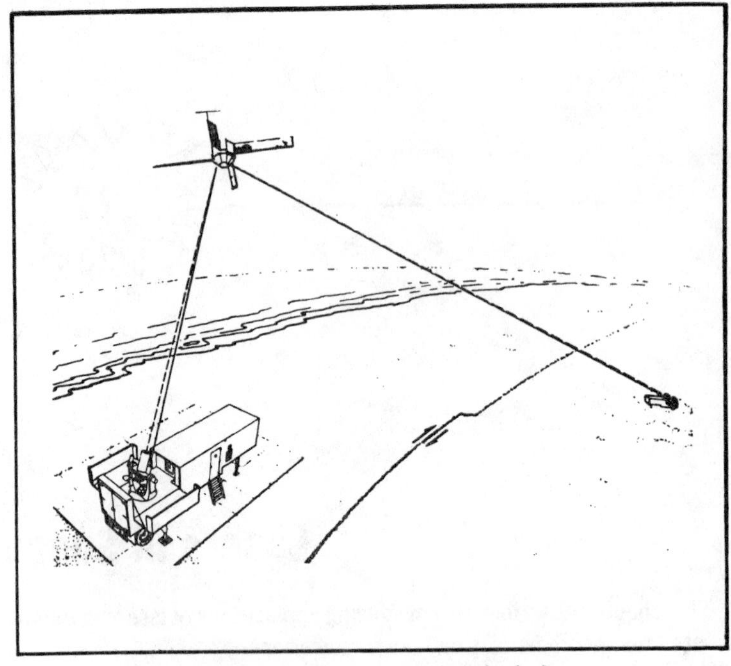

Fig. 17-1. Laser satellite ranging across an earthquake fault.

second, and a pulse width of four nanoseconds. A precision timing system produces a pulse once each second, which initiates the firing of the laser transmitter. A small sample of the transmitted energy is detected by a photodiode. The output pulse from the photodiode is used to trigger a fixed-threshold discriminator, which starts the range time-interval unit. Similarly, the return pulse from the target is detected by a photomultiplier tube, which also triggers a fixed-threshold discriminator, stopping the range time-interval unit. The time between starting and stopping the range time-interval unit is multiplied by the velocity of light to give the range to the target. The computer calculates the azimuth and elevation signals required to drive the telescope mount of the system. It also records and formats the data from each range observation. The reduction of the data is performed at a central computing facility at Goddard Space Flight Center in Maryland. Figure 17-2 shows some typical data after reduction. About 400 points (residuals) are plotted for a 12-minute pass of the satellite. The figure also shows how the precision of the measurements was improved over a 2-year period, from about 50-centimeter precision to about 10-centimeter precision.

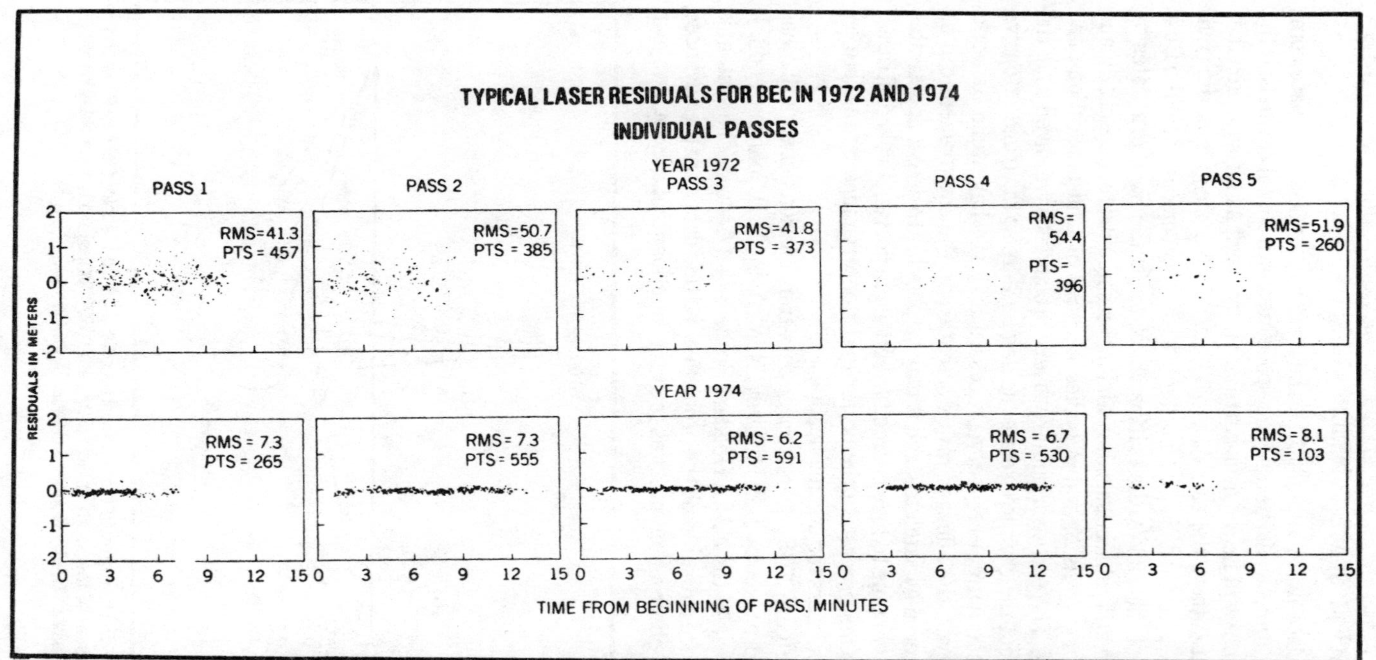

Fig. 17-2. Laser residuals for individual passes of the Beacon Explorer C spacecraft. (Courtesy of the National Aeronautics and Space Administration.)

LASERS IN SPACE COMMUNICATIONS

Once it was thought that the main use for lasers in space communications would be for deep-space missions. Now it seems that the main use will be for near-earth missions, such as for satellites that collect data on the earth's ecology. Man's modification of the environment has had disastrous effects in the past, among them the Dust Bowl of the 1930s, the exhaustion of soil in the Southern states by cotton and tobacco farming, and increasing nitrate, phosphate, and mercury levels in water bodies because of land applications and runoff. The earth's natural recovery processes have saved us in the past, but the changes we are making now are so rapid and extensive that the natural recovery processes may not have time to work. However, the effect of man's activities on the environment can be monitored by satellites, and it may be possible to detect undesirable trends in time to take corrective actions to prevent ecological disasters. One proposed scheme for earth observations by satellite involves lasers for communications.

Figure 17-3 shows a space data-transmission network that has been proposed by J. H. McElroy, N. McAvoy, E. H. Johnson, and J. J. Deganan of the NASA Goddard Space Flight Center. The system is designed to overcome the problem of getting data from the advanced earth-observation satellites to ground stations that are not always in

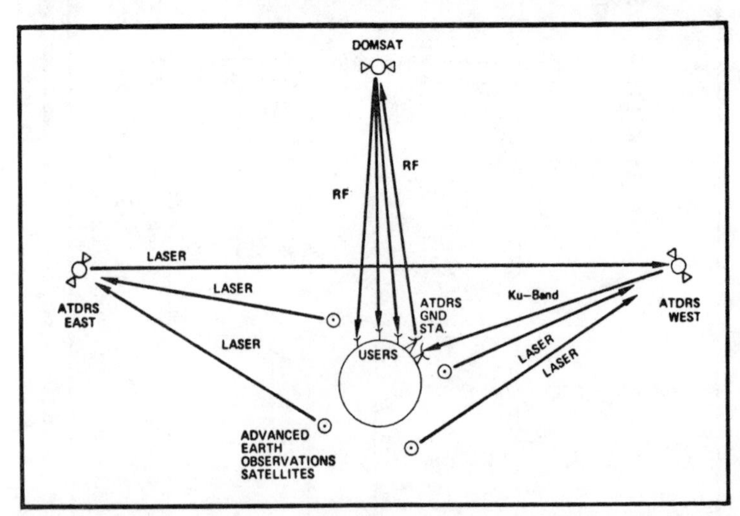

Fig. 17-3. How ground stations will receive data from ecological satellites on the other side of the earth at high data rates (300 Mbps) via a laser relay network. (Courtesy of NASA Goddard Space Flight Center.)

view of the satellites. One solution would be to record the data acquired by a satellite until the satellite is in view of a ground station; however, there would be a reliability problem with tape recorders, particularly at the high data rates involved. So the system in Fig. 17-3 involves a network of relay satellites, called *advanced tracking and data-relay satellites* (ATDRS). These provide for relaying data from the observations satellites to ground as the data is generated. Because of the indirect data paths (such as from observations satellites to ATDRS East to ATDRS West to ground), this is referred to as a *bent-pipe* relay.

A CO_2 laser communications system is used for intersatellite (satellite-to-satellite) communications. The high data rate is again a factor. Laser communication is ideal for high-rate data transmission, which consumes a large chunk of frequency spectrum and can cause frequency-allocation problems. Laser systems also offer freedom from terrestrial radio-frequency interference, and they offer compactness.

In the system of Fig. 17-3, the data from the observations satellites is transmitted to users by an rf relay via a communications satellite (Domsat). Figure 17-3 suggests an application of laser communications for other than environmental data acquisition. When international telephony via satellite must take place over such a great distance that the points of origination and destination are not both within view of a single satellite, the signal must be bounced back and forth between the earth and multiple satellites. This can require as many as three hops and can involve a long one-way propagation time. Also, this mode of transmission uses up valuable frequency allocations for satellite-to-ground links. An intersatellite link by means of lasers would permit two satellites to act essentially as a single system, conserving satellite-ground frequency allocations and reducing propagation time.

Figure 17-4 shows a diagram for a complete receiver-transmitter. The received energy strikes a gimballed, flat mirror (coarse-pointing mirror) and is directed through the telescope, which acts as the receiving antenna. Next is the IMC, or image-motion compensator. This is a piezoelectrically controlled beam steerer that controls the instantaneous field of vision of the telescope so that the telescope tracks the source of the energy (the other satellite). This tracking is necessary because of the relative angular velocity of the two satellites. Without tracking, the laser beam from the source satellite would not intercept the receiver.

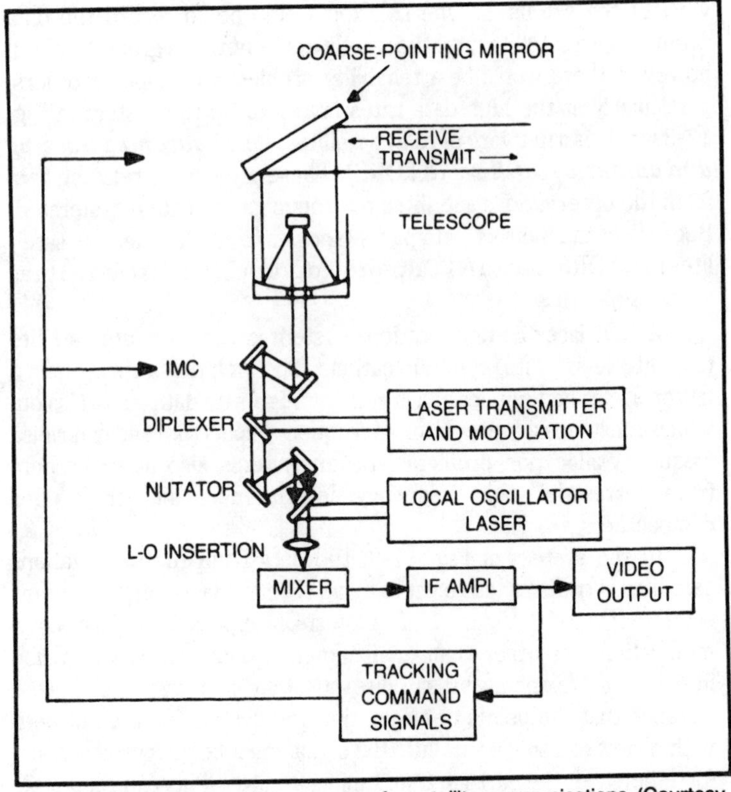

Fig. 17-4. Block diagram of a transceiver for satellite communications. (Courtesy of NASA Goddard Space Flight Center.)

From the IMC the energy passes through the diplexer, which separates the received energy from the transmitted energy. Next is the nutator, which is used to generate a tracking-error signal. Then the local-oscillator laser beam is inserted, and both the received signal and local oscillator beams impinge on the infrared mixer. Like the mixer in any superheterodyne receiver, this generates an intermediate frequency (IF). The intermediate-frequency energy is converted to a video signal.

In the transmitting path the modulated laser energy from the transmitter is diplexed with the receive path and passes through the image-motion converter, telescope, and coarse-pointing mirror, where it leaves the transceiver.

The laser satellite-communications system is still in the experimental stage. A number of problems, such as how to extend the lifetime of space-qualified CO_2 lasers to 50,000 hours, still have to be worked out before the system is launched. When it is launched, a new era of person-to-person communications on earth may be launched with it, and a better understanding of our environment will be assured.

MILITARIZATION OF SPACE

Recently there has been a controversial movement towards the militarization of space. President Ronald Reagan has strongly supported a plan officially named the Strategic Defense Initiative, or SDI. This plan has been more popularly dubbed "Star Wars."

A technical book like this one is scarcely the appropriate place to debate the politics involved. Little technical information on the Star Wars system is publically available, for security reasons. Still, this is an important application for lasers and deserves some discussion here.

Basically, "Star Wars" involves a system of orbiting satellites, armed with directed-energy-beam (laser) weapons. If missiles are detected on a path to the protected area, the SDI satellite will automatically destroy the missile from space, while it is still out of Earth range. The protected area is therefore rendered theoretically invulnerable to nuclear attack.

However, like everything else in politics, nothing is clear-cut, and there is considerable opposition to the plan on a number of grounds. Some doubt the practical feasibility of the system. Can it be made reliable and accurate enough? Can the enormous costs of development be justified?

Such a system can also increase any paranoia on the part of unfriendly governments. If nation A has SDI, but nation B does not, nation A is protected from a first strike from nation B, but nation B would be helpless in the face of a first strike from nation A.

In addition, there is nothing inherent in the technology to prevent a satellite-based directed energy beam weapon from firing upon Earth targets. This is not the intent of SDI, of course, but it is theoretically and technically possible.

As you can see, this is a complicated issue politically. There is a real chance that this nominally purely defensive system could increase, rather than reduce, global tensions.

A directed-energy-beam weapon, such as those in the proposed Star Wars system, must project a laser beam of destructive force for a considerable distance. A very powerful laser device must be used. Moreover, for practical use on a satellite, it must be relatively small and lightweight. The Department of Defense has sponsored considerable research into free-electron lasers, which are high-energy/ high-current devices operating in the millimeter range.

Free-electron lasers have numerous other potential military (and some civilian) applications, in addition to Star Wars. These include compact radar telecommunications transmitters and the jamming of enemy communication systems.

Lasers in Industry

The lasers of interest in material-processing applications are the ruby, neodymium-doped glass (Nd:glass), Nd-doped YAG (a variety of garnet crystals, yttrium aluminum garnet) and carbon dioxide (CO_2) types.

In order to be a preferred solution for a particular material-processing application, the laser machine must have advantages over alternative methods on the basis of the following factors: cost comparison, convenience, performance, and operational requirements. Material-processing applications can generally be broken down into the following categories: hole drilling, welding, material removal, cutting, scribing, sealing, and trimming.

Manufacturers offer as standard products complete industrial systems for resistor trimming, gyro balancing, silicon cutting, sealing, trimming, drilling, material removal, and welding. The following section describes the more common types of laser applications.

LASER OPERATION

The lasers most commonly used for both drilling and welding are ruby, Nd:glass, Nd:YAG, and CO_2.

Pulse widths that are optimum for drilling depend on the hole desired and the material, but are in the general range of 100 to 1000 microseconds. Shorter pulses are wasteful of energy because the evaporated material must have time to escape in order to prevent absorption of the incoming radiation. Longer pulses lose energy through conduction and radiation. Welding pulse widths are typically 0.5 to 20 milliseconds, the most commonly used range being two to six milliseconds.

Ruby has the advantage of being capable of high average powers at moderate to high pulse energies, typically 10 to 20 watts average, with pulse energies on the order of 5 to 40 joules. The ruby is normally operated at pulse rates less than one pulse per second to avoid excessive heating of the rod, which would result in a change in output energy and beam divergence. Such changes are unacceptable for drilling and welding unless a constant repetition rate is used with a series of warmup shots to produce equilibrium. Only in the case of truly *continuous* production welding is such a constant-repetition-rate system practical.

Nd:glass has the advantage of higher efficiency, with output energies equaling or exceeding those from ruby. The low thermal conductivity of glass presents a problem in systems where a repetition rate of greater than about ten pulses per minute is required, but constant repetition rate with a warm up phase permits operation at higher pulse rates.

Nd:YAG lasers under CW operation are applicable (a) to drilling when Q-switched, producing a series of short pulses which chip away at the material at a rate of 100 to 10,000 pulses per second (this is extremely effective for many applications), and (b) to seam welding. Power levels of hundreds of watts are available.

CO_2 lasers feature high average power output and efficiency. These lasers are used for cutting or welding when operated continuously (CW) and will drill many low-thermal-conductivity materials in this mode. By pulsing the discharge repetitively, multijoule pulses can be obtained at a rate of about 100 per second. For drilling through metal this type uses short-duration (a few microseconds) pulses to remove material before the laser heat can flow away, and the high energy available overcomes losses due to the high reflectivity of metals at 10.6 microns.

HOLE DRILLING

The high intensity of a laser pulse delivered in a very short interval will cause materials to vaporize, thus creating a hole. By focusing the laser beam through an optical system, the energy output can be delivered into a very small spot size, typically from 0.005 to 0.050 inch. Using a pulsed solid-state laser with four to ten kilowatt peak-power output, holes can be drilled in many hard-to-machine substances, such as alumina ceramic, tungsten, tool steel, tantalum, and niobium.

Laser drilling can be accomplished, where required, in inert-gas-filled chambers, because there is no need to be in physical contact with the work. By changing the laser output parameters, cylindrical or conical holes can be produced.

There are several processes which occur during a drilling pulse: initial heating of the surface (occurring at typical rates of 10^{10} degrees per second), vaporization of surface material, and penetration of the beam into the material. Evaporated material of high-density accumulates in the hole, melted material washes from the walls, and pressure inside the hole reaches 10^2 to 10^3 atmospheres. A jet then flows out at supersonic velocity, and further erosion of the hole occurs due to the hot vapor (this is sublimation). Because of this complex process, developments of drilling techniques involve experiments; however, general guidelines exist:

1. The laser energy required to remove an amount of material is aproximately equal to the ***sublimation energy,*** the energy required to turn a material from a solid into a gas once a threshold energy density is exceeded.
2. The threshold energy for many metals is approximately 10^4 joules per square centimeter.

The light-pipe effect (reflections of the laser beam from the hole walls) has yielded very large depth-to-width ratios, even with short-focal-length lenses. It is possible, for example, to drill holes in high-melting-point metals, such as Rene 41, with diameters of 0.020 inch and a depth of 0.120 inch at less than 1° taper, with a single shot from a ruby laser. Holes two to ten microns in diameter with a depth 250 times the diameter have been obtained with a repetitively *Q*-switched CW Nd:YAG laser. A CW CO_2 laser can drill holes through plastic with very high ratios of depth to diameter, drilling at such a slow rate that the action is clearly observable.

A fairly typical industrial laser unit is the Raytheon Model SS-347 YAG diamond-die driller, which works by generating laser energy pulses that, when focused on the diamond surface, cause the vaporization of the diamond in small quantities. The laser is fired at high pulsing rates, and the die is slowly rotated during the firing. These actions create extensive overlapping in order to achieve hole roundness. The natural conical shape of the focused beam provides the desired bell-shaped entrance into the diamond die.

Viewing of the drilling process is monitored by a closed-circuit

TV system. A magnified image of the die being drilled is reproduced on the TV screen for complete control. The top side is illuminated for focusing onto the diamond surface, and the bottom side is illuminated for detecting breakthrough and for sizing the hole. A unique feature of the optical system, called *coplanar focusing*, provides the necessary focal point for both viewing and drilling.

WELDING

Laser welding uses lower peak powers than those used for drilling. To prevent loss of material through vaporization, peak power must be kept low. To cause a weld to form, both pieces must absorb enough power to become molten and fuse together. This usually requires high average power. The best welding results are accomplished using wide (i.e., three- to eight-millisecond) laser pulses having low peak but high average power.

Welding or bonding is an example of a field in which there can be very wide use of lasers. There are a number of possible advantages of a laser over other techniques: noncontact of dissimilar materials, precision control, small-area coverage, access to normally inaccessible areas, transparency to glass, a small heat-affected zone, and no vacuum requirement.

Both spot and seam welding can be performed with lasers. Metals, plastics, glasses, and ceramics have all been successfully bonded. Rather than list a lot of examples of successful laser welding, a few cases will be examined in more detail.

Beam-Lead Welding

Techniques have been developed for welding beam-lead integrated circuits to thin-film interconnecting circuits on a substrate. One laser pulse per lead weld is impractical because of the time required, excessive manipulation of the circuit or laser beam, and the overall cost per weld. A configuration in which a rectangular, focused laser-beam pattern is produced used mitred cylindrical lens sections. This pattern is created with a uniform energy density by using a diffusing element which reduces variations of beam intensity at the lens entrance. In this way, we can obtain good welds on a 16-lead device in one shot. Special holddown techniques that use a transparent Teflon membrane and a vacuum plate are essential to reproducible bonding.

Seam Welding

Seam welding can be performed either by overlapping spot welds or by using a continuous (CW) laser. The analysis of continuous welding, including the balance between welding and cutting, is very similar to the pulse case. The CO_2 laser has been used to weld plastics. The high CW output power from Nd:YAG lasers promises wide application of this source for seam welding. The absorption of radiation by metals is much higher at 1.06 microns (Nd:YAG) than it is at 10.6 microns (CO_2). In addition, the convenience and lower cost of conventional glass optics usable with Nd:YAG is a great advantage over special infrared transmitting materials, such as germanium, required for CO_2.

The availability of substantially more power at higher pulse rates and higher efficiency makes the modern pulsed YAG laser an excellent candidate for production applications requiring precision welding. Particular advantages are: very high precision, very low thermal damage, capability for welding extremely hard metals, high welding efficiency, high process rates, and compactness of a laser system.

Precision Metal Removal

This type of work is similar to drilling in that material is vaporized. An excellent application for precision metal removal is gyro balancing. Gyroscopes run at very high speeds, typically 10,000 to 25,000 rpm, and must be balanced to rigid specifications. Conventional gyrobalancing methods require placing the gyro in a test jig and running up to operating speed (5 to 20 minutes). Then, imbalance is sensed with a strobe sensing device. Before any metal can be removed, the gyro must be allowed to run down and must be stopped without any damage (10 to 20 minutes). A skilled technician then securely clamps down the gyro (to prevent bearing damage) and removes a small amount of metal at the imbalance point (5 to 15 minutes). The gyro is then returned to test jib, and the process is repeated until the gyro is completely balanced (3 to 5 hours).

Using the laser balancer, time required is cut to approximately 10 to 15 minutes. This is accomplished by using a 12-inch Nd:glass laser head which gives a short pulse (approximately 50 microseconds) at up to 10 joules, which will vaporize metal. The gyro is run up to operating speed and the imbalance sensed by conventional methods. The imbalance sensor sends a signal to the laser electronics representing the imbalance point on the gyro. The laser electronics then computes when the laser should be fired to obtain maximum metal removal at the imbalance point, then sends the fire signal to the laser head.

The laser gyro-balancing technique does not require removing the gyro mechanism from the machine. Gyro balancing is accomplished in one continuous operation at a substantial time saving while achieving better balancing.

CUTTING

Continuous-wave CO_2 or Nd:YAG lasers are excellent for use in cutting operations. Advantages of laser cutting include: no physical contact between laser cutting tool and work (therefore, contamination of workpiece is reduced or eliminated); thin (0.01-inch) line width of cut, operation can be carried out in enclosed vessel; vessel can be gas filled; and cuts are sterilized.

Since most plastics (even transparent types) are opaque to the 10.6-micron wavelength of the CO_2 laser, these materials can be easily cut with this type of laser.

The Q-switched YAG lasers offer high peak powers for cutting thin pieces of most materials. Since the Nd:YAG laser's output wavelength (1.06 microns) differs by a factor of ten from that of CO_2, narrower cut widths are possible.

The addition of an oxygen jet can produce much deeper cuts at considerably faster rates in many materials due to an exothermic reaction.

SCRIBING

Scribing is similar to cutting in that some material is removed from the workpiece. Materials otherwise hard to scribe—such as sapphire and most crystalling substances—can be scribed prior to breaking.

Laser scribing can be used advantageously in the microelectronics field with the following advantages: narrow scribe width; can be carried out in an inert atmosphere; scribes are sterilized and contamination free; and repeatable results.

SEALING

Plastics can be sealed with a CW CO_2 gas laser quickly while maintaining sterility. Seals are similar to those obtained by conventional methods, but more variations in patterns can be obtained with less retooling using laser techniques.

An example of a modern high-power metal-working machine is the Model 900 of the Holobeam subsidiary of Control Laser Corporation. It incorporates a high-power continuous YAG laser, which is mounted on a Bridgeport milling machine. The laser is available in powers of 200, 400, and 600 watts. A numerical control unit is used to guide the laser through the desired contouring motions. The machine is used to make welds and cuts. It can also be used to scribe, as in inscribing serial numbers. The laser operates at 1.06 microns, which is a very efficient wavelength for metal working.

Holobeam's Model 990 is a welder, driller, and pressurizing system. The workpiece is drilled (holes range from 0.002 to 0.020 inch) through a window in the pressure chamber. An inert gas is then forced into the pressure chamber and thus into the drilled hole. When the pressure inside the workpiece reaches equilibrium with the chamber pressure, the hole is closed through the window with a welding pulse from the 990 laser. This entire process takes less than 30 seconds. The advantages of the unit include no drill bits to wear out, ability to drill repeated holes in a vacuum, extremely fast operation, elimination of contamination, capability for automatic programming, and capability for both welding and drilling in a single unit.

Figure 18-1 shows the construction of the laser head for a Holobeam continuous-wave Nd:YAG laser. This type of laser is used for measuring applications, such as air-pollution monitoring, and for industrial applications, such as resistor and watch crystal trimming, silicon and ceramic scribing, and diamond and ruby drilling. These lasers have a transparent plastic laser head. A single krypton-arc pump lamp is used, which is easily replaced. Safety interlocks ensure safe operation. The power supply is interlocked to the laser head water-cooling system. A cutoff safety shutter built into the spatial filter ensures instant beam cutoff when activated. A relative-power monitor oversees CW or Q-switched operation and is connected to the remote-control console. All laser operations can be monitored and controlled from this console.

In Q-switching, an electro-optical shutter, such as a Kerr cell, is interposed between the active laser material (ruby, etc.) and the partially reflective mirror of the system. The shutter holds back the laser's energy somewhat as a dam holds back water. When the shutter is opened by applying a voltage to it, the laser light is suddenly released as a flood of energy. This technique produces pulses of hundreds of megawatts for a few millionths of a second. Under some

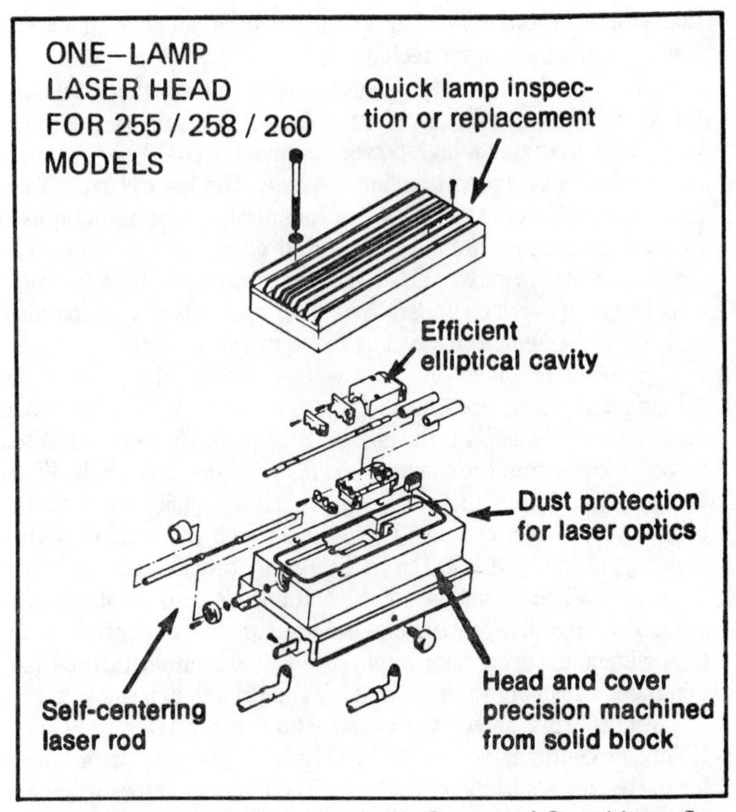

Fig. 18-1. Construction of the laser head. (Courtesy of Control Laser Corporation.)

circumstances, in some materials, 150 watts of pulsed YAG power has greater penetrating power and welding efficiency than 400 watts of continuous YAG power.

A pulsed laser than can be used in various scientific and industrial applications is shown in Fig. 18-2. Both Q-switched and conventional-mode ruby, Nd:glass, and Nd:YAG lasers are available with pulse energies ranging up to 150 joules and peak powers of over one gigawatt (1000 megawatts). The repeatability of the pulses is aided by an integrated power supply and cooling unit, which is controlled from a remote console. A massive U-channel rail provides long-term alignment stability. Other design features include a free-floating lamp mount, a self-aligning cavity, and O-ring water seals between the rod and the cooling system.

IMPROVING EFFICIENCY

While laser machining offers some significant advantages over more traditional industrial methods, present laser techniques for cutting and milling metals tend to be rather slow, and eat up a lot of energy.

Some researchers at the Massachusetts Institute of Technology suggest that this may be because the laser systems are too closely modelled after traditional industrial tools. The laser is a unique device, and should be treated uniquely to maximize efficiency.

Today a laser is generally used to remove unwanted metal by vaporizing successive surface layers. This technique is modeled after the way a mechanical head chews away the unwanted material. This technique certainly works—that's why it is commonly used. But it may not be the most time- and cost-efficient approach.

The MIT researchers suggest taking advantage of the laser-unique characteristics in the job, and carve off chunks of unwanted material.

The new proposed method calls for splitting the laser beam into a pair of sub-beams with a mirror. Then both of the sub-beams are aimed at a common point, but from differing angles.

To see how this works, let's consider an application requiring a

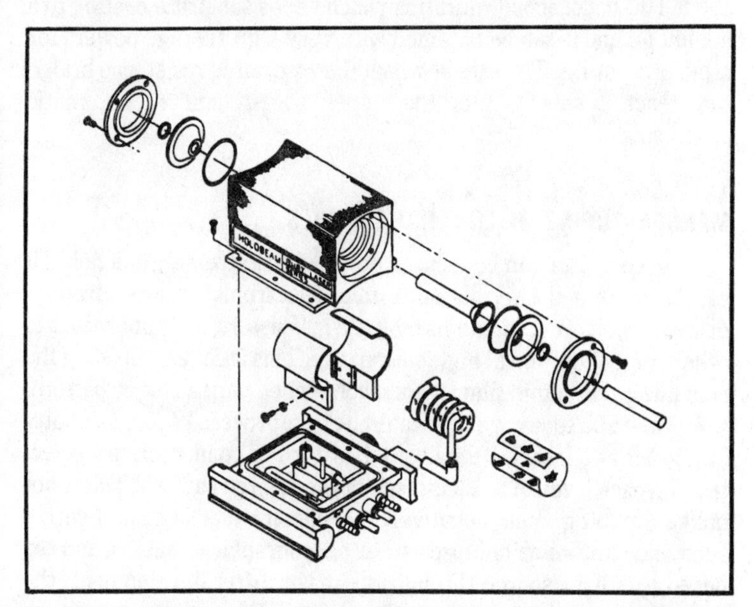

Fig. 18-2. A pulsed laser for scientific and industrial applications. (Courtesy of Control Laser Corporation.)

metal rod to be reduced in diameter. One sub-beam would be positioned to come in from the end, while the other sub-beam would be aimed into the curved surface of the rod at a 90° angle. While the laser is projecting these split beams, the rod is rotated in a lathe. The intersecting sub-beams will neatly cut off ring-shaped sections from the rod, reducing its diameter.

RESISTOR TRIMMING

Use of a laser for resistor trimming offers several advantages: little or no residue to contaminate the circuits, negligible heating of the surrounding components, control that is immediate and precise, and a spot size much smaller than that obtainable by mechanical methods.

The types of lasers that are recommended for this application are: (1) CO_2, both repetitively Q-switched and repetitively pulse-pumped, and (2) Nd:YAG, CW pumped and repetitively Q-switched.

The CO_2 has lower basic cost, with the disadvantages of requiring optics of more exotic materials and of producing a relatively large spot size as a result of the long wavelength of the laser light.

The CW pumped, Q-switched Nd:YAG provides a short pulse of about 100 nanoseconds duration which keeps substrate heating to a minimum, and it can be obtained with very high average powers for rapid processing. The rate at which the associated resistance bridge can react currently sets the upper speed limit of automatic processing.

MANUFACTURING INTEGRATED CIRCUITS

Since a laser can be focused down to an extremely fine point, it can be used in the formation of microelectronic circuits. An integrated circuit, or IC, contains multiple transistors, and simulations of other components on a single silicon chip. This is done by dividing the chip into several thin-film layers, each etched with a unique pattern.

Most ICs today are fabricated using a process known as photolithography to inscribe the desired pattern of conductors or dielectrics on each layer. A mask is made of the desired pattern. This is not unlike a photographic negative. The silicon wafer is coated with a substance known as photoresist, or photographic emulsion, and exposed to a light source through the mask. After development, the desired pattern is etched into the wafer. Features as small as one micrometer can be created using this method.

There are several ways to grow insulating or conducting films. These include condensation of vaporized metals, oxidation of the silicon substrate, and gas-phase chemical reactions.

Etching of the films can also be done with liquid chemical baths (acids), or with gases excited by an electrical discharge (plasma etching).

Recent developments have taken advantage of the unique capabilities of the laser. A laser can be used to either burn away undesired material, or to effect chemical reactions at the surface level. The laser can be tightly focused to limit the range of the effects. In fact, a laser beam can be focused down to a spot comparable in size to the wavelength of the laser light. Even a low-power laser can produce a high-intensity beam when focused down to micrometer dimensions.

A laser may be employed to create localized thermal reactions, such as the creation of thin-film resistors. But a properly focused beam can also be used to create a localized chemical reaction on the wafer's surface. Thus a laser may be used for etching, doping, or deposition. In each of these processes, the laser beam is focused onto the substrate, which is then moved in a plane perpendicular to the beam's axis to write the desired patterns onto the semiconductor wafer.

Another way lasers may be used to create fine patterns is interferometric imaging. This takes advantage of the laser's high temporal coherence (monochromaticity). If a laser beam is split into a pair of sub-beams which are then recombined at the substrate's surface, extremely fine interference patterns can be created. A special advantage of this technique is that no photolithographic mask is required.

19

Lasers in Commerce

A number of recent developments in laser technology portend big changes in the world of commerce. Although still in the prenatal stage of development, these techniques may soon be delivered from the research laboratory into the real world of computers, image and data displays, and printing.

ALTERNATIVE TO CRT COMPUTER DISPLAYS

In a paper delivered at a meeting of the Society for Information Display, six scientists from the San Jose Research Laboratory of International Business Machines Corporation described an experimental terminal built to demonstrate a new information-display technology. It uses a deflectable laser beam to write alphabetic and numeric characters in a cell containing a specially formulated liquid-crystal material. The laser-written characters are projected onto the rear of a translucent display screen for viewing at 25 times their original size. With further developments, the new technology may be useful as an alternative to the cathode-ray-tube displays used in computer terminals. The characters are more sharply defined than in CRT displays and appear black on a white background, giving the display the easy-to-read look of a typewritten page.

A key element of the new technology is a small but powerful solid-state laser that operates at room temperature, producing up to 25 milliwatts of power in its continuous-wave output beam. This laser is a chip only 200 by 400 microns. The diameter of a human hair, by comparison, is about 100 microns.

Gallium-arsenide laser systems are being developed for this display and other applications at IBM's Thomas J. Watson Research Center in Yorktown Heights, New York.

To write characters, the laser beam is focused onto a translucent liquid-crystal material sealed in a flat glass cell about the size of a photographic slide. Two computer-controlled oscillating mirrors deflect the beam across a portion of the cell, producing a scanning motion in the horizontal and vertical directions. The beam is turned on and off at the right times to form characters from patterns of overlapping spots in a grid nine spots high by seven spots wide.

The display terminal uses lenses, mirrors, and a projection bulb to project reflected images of the characters being formed in the liquid crystal onto the rear of a viewing screen. The magnified character images, originally 90 microns high, appear on the screen about 2.25 millimeters high. This is nearly a tenth of an inch, or about the size of the character produced by a typewriter.

Color slides can be inserted into the terminal and projected onto the screen as overlays to the material being typed at the terminal keyboard. With the use of such overlays, the entering and updating of material is a simple matter of filling in predetermined forms.

Spots formed in the liquid crystal are caused by heat from the laser beam. This heat upsets the alignment of molecules in a small region of the material, and the misaligned molecules scatter the projection light, producing a dark spot in the magnified image on the screen.

Normally, the information written in the liquid crystal will remain indefinitely because of the material's high viscosity. But the display can be erased faster than a person can perceive by applying an audio-frequency voltage across the plates that confine the liquid crystal. Moreover, portions of the information can be erased by reducing this voltage to less than a critical value and reheating only the selected proton of the liquid crystal with the laser beam.

Writing of characters is done only when the beam is scanning from left to right and from top to bottom within a single character location. With this method of scanning, it takes the laser 50 milliseconds to form a character—a rate of 20 characters per second.

Although the speed is sufficient for entering information from a keyboard, it is much too slow for practical use in displaying information that is already stored in a computer. IBM is exploring ways to increase the speed of image formation. These methods include using lasers with higher output power, improving the optical system to focus a larger percentage of the laser output onto the liquid-crystal cell, developing liquid crystals with greater sensitivity to light, and using arrays of lasers to write more than one row of spots at a time.

INTEGRATED LASER AND FIBER-OPTIC DISPLAY

In a development that may have significant implications for the liquid-crystal-display (LCD) terminal just described, researchers at IBM's Thomas J. Watson Research Center have created an integrated package containing all the electro-optical elements of a fiber-optic transmitter. The package, as illustrated in Fig. 19-1, contains a semiconductor laser array, a cylindrical lens, and an array of fiber-optic light guides. These components are mounted on a silicon wafer, which also contains thin-film drive electrodes for the lasers.

The chief advantage of the silicon mount is that it serves as a high-precision, miniature optical bench for the optical elements. Using photolithographic techniques and a special (anisotropic) etch; grooves are produced whose shape and depth are controlled precisely by crystalline planes in the silicon. The optical elements are placed in these grooves, which automatically align the components to the optical axis of the package.

This very simple packaging technique couples up to 70% of the light from the lasers onto the optical fibers. Excellent optical effi-

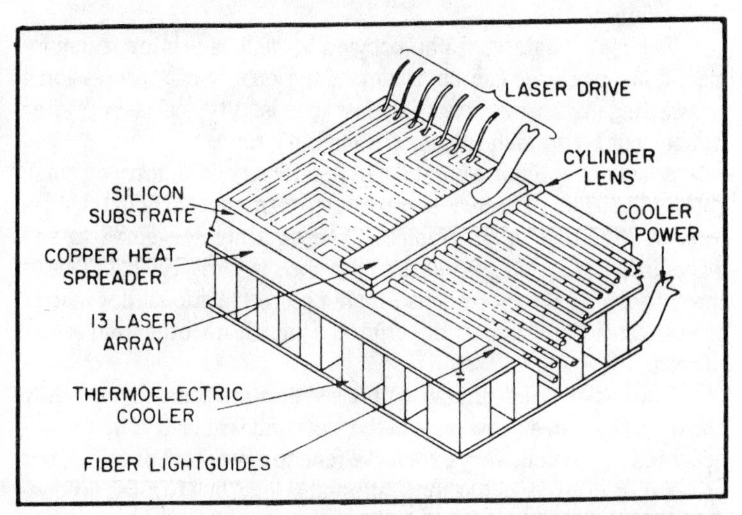

Fig. 19-1. A schematic drawing of the integrated fiber-optic transmitter discussed in the text. (Courtesy of IBM Corporation.)

ciency is thus attained with a fabrication process that offers high potential for integration.

The silicon chip also provides electrical isolation between devices, good heat dissipation, and a smooth surface for laser-array bonding. Future packages may also integrate active electronic devices into the chip. The package thus gives an optimum set of characteristics to make gallium-arsenide lasers practical for electro-optic applications in display, printing, and communication.

The laser arrays used contain up to 13 lasers, fabricated in a single bar of gallium-arsenide — gallium-aluminum-arsenide double-heterojunction material. The individual lasers have an output power of 50 milliwatts in continuous operation. This relatively high power level is needed for experiments on a variety of possible electro-optical applications besides communications.

The silicon optical bench is laminated to a copper heat spreader, which, in turn, is mounted on a thermoelectric cooler. The junction temperature of the lasers is maintained at about 30°C.

The cylindrical lens used to focus the divergent output light from the lasers into the fiber light guides is itself a short piece of glass fiber with a diameter of 70 microns.

LASER PRINTERS

Lasers can also be used to provide permanent hard-copies of computer data. One of the latest "hot items" on the market is the laser printer.

Taking advantage of photocopier, as well as printer technologies, a laser printer can approximate the quality of a professional typesetting machine at amazingly fast speeds. A typical laser printer can put out six to eight pages in a minute's time.

A laser fires minute bursts of high-intensity light, forming small dots on the paper. These dots combine to form the individual characters, or high-resolution graphics. The dots are too small to see individually. A rating of 300 dots per inch is fairly typical. This is three to five times more dense than a more traditional dot-matrix printer can accomplish, so the printout is much smoother and easier to read.

The major disadvantage with laser printers today is that they are very expensive. However, in the electronics industry, new products tend to start out very expensive (due to development costs), but prices soon drop down to a more affordable level due to mass production and increased competition.

A NEW KIND OF PHOTOGRAPHY

In a paper published in "Applied Physics Letters," IBM scientists M. D. Shattuck, R. V. Pole, and G. T. Sincerbox have reported the discovery of a new approach to recording images. Like ordinary photography, it captures a picture by exposing a film to light. But instead of using photography's chemically triggered and irreversible development process, it initiates the development of latent images by switching on an electrical field.

The new image-recording approach is based on an electrochemical phenomenon called *photoinduced electrochromism* (PIE). The IBM scientists observed this phenomenon in a transparent liquid film containing molecules of a class known as *pyrazoline*. Exposure of the liquid to a laser beam produces, they found, a faint blue-green image that becomes darker when a developing field is turned on. The solution retains the color even after the field is removed. If the direction of the field is subsequently reversed, however, the developed image disappears. This makes the recording solution available for many cycles of reuse.

Electrochromism, which itself is not a new discovery, is the ability of certain materials to change colors under the influence of an electric field. The IBM discovery represents the first time that a reversible electrochromic effect has been induced by light.

The experimental cell in which the PIE effect was observed was formed by sealing the pyrazoline-containing liquid between two pieces of glass that were coated with a transparent layer of electrically conductive material. After illumination of the cell with blue laser light, a faint blue-green coloration was observed in the irradiated areas of the liquid. When a dc voltage was then applied across the conductive surfaces of the cell, the colored areas quickly darkened by an amount proportional to the applied voltage and to the duration of its application. The developed image persisted in the experimental cell for several minutes after the voltage was removed.

The researchers then found that reversing the voltage across a cell containing a fully developed image started a bleaching process that erased the image in less than a second. In early experiments with PIE the researchers achieved several dozen cycles of exposure, development, and erasure before the cell became degraded enough to require the preparation of a fresh solution.

The image formed in the IBM experiments was a two- by two-inch line pattern commonly used for testing the resolution and reproduction quality of photographic equipment. The pattern was formed by turning a scanning laser beam on and off at points corresponding to the beginning and end of lines. A resolution of about 300 lines per inch was achieved.

The energy density required to expose an image in a PIE cell is on the order of ten microjoules per square centimeter, an amount comparable to that used in exposing a medium-speed photographic film.

The possibility of recording information with a finely focused deflectable laser beam gives the PIE approach a higher image-forming resolution than matrix-type electrochromic displays in which picture elements can appear only at points in a predetermined grid.

Another type of laser photography is holography. By simultaneously photographing an object from multiple angles, a three-dimensional image can be recorded. The photographed object can be looked at from any angle.

Lasers are required both to take and to view holographs. The 3-D effect can be stunningly realistic.

Currently holographic applications are rather limited due to the complexity and expense of the required equipment, but research is continuing, and three-dimensional holographs are likely to become increasingly common in the near future.

PROJECTION TV BY LASER

Large-screen projection televisions are popular, both in public places, such as bars, and in the homes of individuals who want the full theatrical experience when watching movies. Unfortunately, image brightness can be a problem. Many of the projection TVs now on the market require the viewer to minimize room lighting, at least to some extent.

An English company, Dwight Cavendish Displays has developed the Spectrolight projection TV, which uses lasers to project the image. Since laser light is inherently intense, there is no problem with normal room lighting. The projected image can be up to 50 feet wide.

Red, green, and blue laser beams are passed through a modulating crystal for the desired hue and luminance information for the picture. A rotating mirror is then used to scan the laser light to separate the consecutive lines used in traditional video systems. These lines are focused onto an oscillating mirror that projects the image to the screen.

While intriguing, this approach is not quite ready for a consumer product. For one thing, few of us have living rooms large enough to accommodate the equipment. An even more significant problem is that this is a very high energy unit. It consumes a lot of power—about 28 kilowatts, which is likely to have a major impact on anyone's electricity bill. And with that much power being consumed, there is a lot of waste heat generated. The system must be water cooled. About 4.75 gallons are required per minute.

Still, the technology looks promising, and in the future we may well see practical laser-based projection TVs.

RECORDING DIGITAL DATA VIA LASER

One of the most influential products of recent years is the compact disc. Most people have undoubtedly seen audio CDs, which offer crystal-clear sound. But this is an invention of interest to more than just the high-fidelity enthusiast.

A compact disc, or CD is a polycarbonate plastic disc about 12 centimeters (4.72 inches) in diameter, with a 15 millimeter hole in the center for the drive spindle. The disc is 1.2 millimeters thick.

Digital data is encoded on a CD as a series of indentations and flat areas known as "pits" and "lands." This is illustrated in Fig. 19-2.

When a CD is manufactured, the first step is to make a master disc, usually out of glass. A laser beam is used to burn the appropriate pits into the glass disc. The data is recorded in a spiral that moves from the center of the disc to the outside rim. This is the opposite of a standard record album.

The completed glass master disc is used to mass produce the actual CDs. An injection-molding process is employed. The data substrate in the final CDs is a polycarbonate plastic. A protective

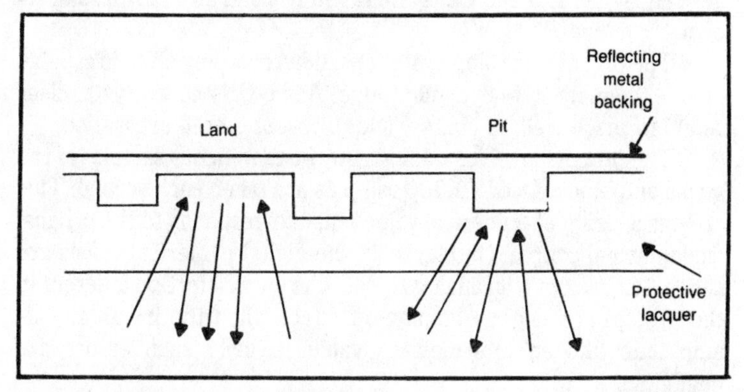

Fig. 19-2. Digital data is encoded on a CD as a series of "pits" and "lands."

plastic coating is placed over the data surface. This coating is transparent, of course. The back side of the disc is covered with a reflective layer. Finally, a label and a protective lacquer are added, and we have a completed compact disc.

A laser is also used in the playback process. An oval laser beam is generated by a gallium arsenide laser. A collimating lens, which makes light rays parallel, converts the oval beam into a circular shape. The next lens changes the beam's shape again, this time into a sharply convergent cone. Finally, the optical head positions the tip of the beam cone over the rotating data track on the disc. The entire process is illustrated in Fig. 19-3.

The light is reflected back from the data track. The amount of light reflected back depends on whether the beam is hitting a land or a pit.

The reflected light is picked up by a photodetector and is converted into a proportional current. Additional circuitry decodes the fluctuating current into a string of digital pulses.

Audio CDs

Currently, the most wide-spread application for the compact disc is in audio. The digital laser player offers a number of advantages in a high-fidelity audio system. The discs are virtually indestructible. They can be damaged with severe misuse, but when compared to standard vinyl records, which can easily be scratched, or magnetic tape, which can break or stretch, CDs are exceedingly durable.

CDs are nominally free from the hiss that plagues all analog recording media. Of course, any CD made from an analog original master tape will reproduce the hiss on the source tape along with the desired signal. But the CD system will not add any new hiss of its own.

The digital recording system used in recording CDs, is capable of an extremely wide dynamic range. Audio CDs have crystal-clear sound (if made well, of course) and excellent stereo separation.

The process involves taking many instantaneous samples of the signal amplitude. Over 40,000 samples are taken each second. This creates a string of numerical values that corresponds to the original analog sound source. On playback, the digital numerical values are converted back into an analog signal. The only distortion inherent in the system is a slight rounding-off effect when the instantaneous amplitude hits an intermediate value. Usually such errors are negligible.

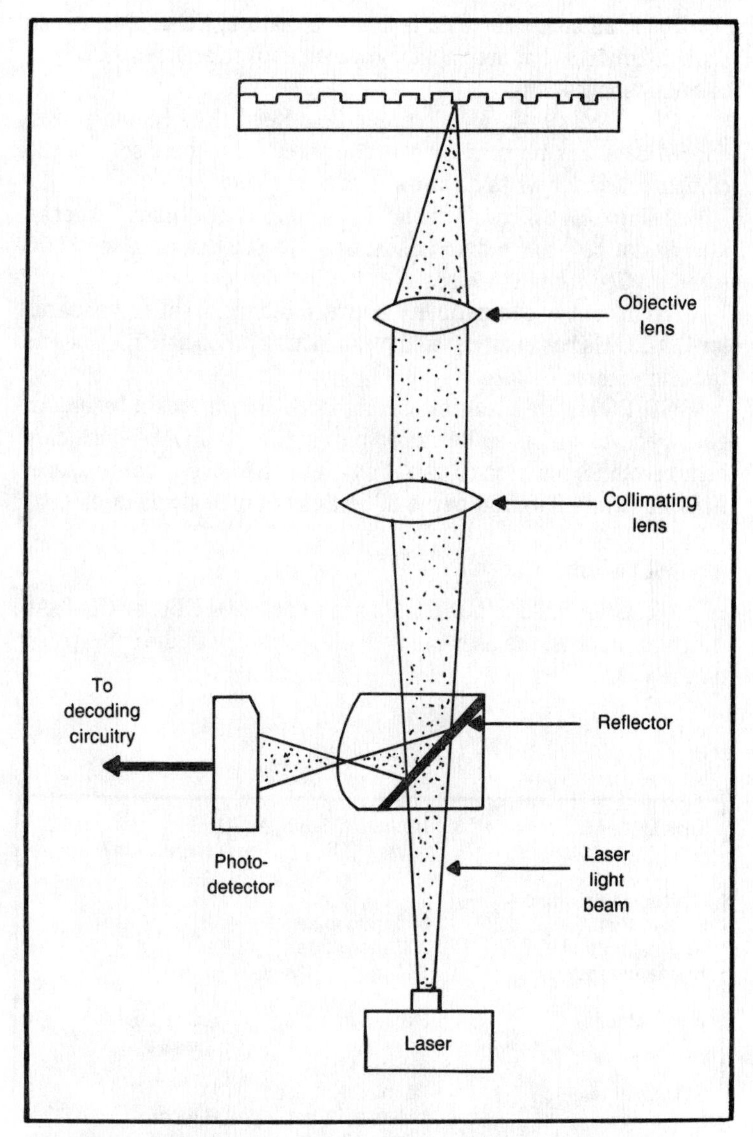

Fig. 19-3. How the laser beam is focused to read data from a CD.

CD-ROMs

While audio recording is currently the most popular application for laser-read compact discs, this recording medium can be used for permanent storage of any data that can be encoded in digital form.

Since all computer data is digital in nature, CD-ROMs are an obvious product that are rapidly growing in popularity, especially in business applications.

CD-ROM stands for Compact Disc-Read Only Memory. Recorded data can be read by the computer from the disc, but the computer cannot write any new data to the disc.

A single 4.72 inch CD can hold 550 megabytes of data. To get an idea of just how much data this is, one CD-ROM can replace 1500 standard $5\frac{1}{4}$-inch floppy discs.

To maximize compatibility between brands, a set of standards for CD-ROMs has been devised by an industry association known as the High-Sierra Group.

CD-ROMs are ideal for storing large constant data bases. An entire encyclopedia can be stored on a single CD-ROM—including high-resolution graphics illustrations, and still have room to spare. Approximately 250,000 pages of text can fit on a single disc.

Interactive CDs

Closely related to CD-ROMs are Interactive CDs, or CD-I. As the name implies, the user can actively interact with the data stored on the CD.

Table 19-1. CD/I Modes.

Text Modes	Maximum Storage	User Manipulation?
Character-encoding system test	600 megabytes	Yes
Application text	300 megabytes	Yes
Bit-mapped text	120 megabytes	No
Audio Modes	**Maximum Time**	**Fidelity Level (Equivalent)**
	Stereo Mono	
CD-digital audio	72 min. Not used	CD
Hi-fi music	144 min. 5 hrs.	LP record
Mid-fi music	5 hrs. 10 hrs.	FM radio
Speech	10 hrs. 20 hrs.	AM radio
Video Modes	**Colors**	**Maximum Frames Per Disc**
"Natural" pictures	- - - - -	550
High-resolution graphics	256	60000
User-manipulated graphics	256	5500
User-manipulated graphics	32,768	2750

Data in a CD-I system can be in any of several modes, as outlined in Table 19-1. These modes can be combined on a single disc, so the system is incredibly flexible.

The applications for CD-I are almost limitless. Self-directed education programs and games are sure to be popular. An automotive CD-I system could contain road maps and act as an automated navigator. This approach is also being employed in ships at sea, with oceanographic data.

CD-I can also be used for armchair "travel." Thousands of photographs from an area can be stored on a disc. The user can then wander about the area as he likes, by using the system controls to decide what picture will be displayed next.

In concept, CD-I is not really dissimilar to CD-ROM, and can be used for many of the same types of applications. The main difference is that a CD-ROM drive is designed to be used with an existing computer system. A CD-I player is self-contained.

Video Discs

A similar system to CDs is used in laser video discs. The discs are somewhat larger than CDs, and the program is recorded as analog rather than digital data. Full, natural animation in a digital format would take far too much space (a few thousand bits per frame).

Laser video discs offer much sharper images and better sound quality than any of the consumer video tape formats. However, the discs are not user recordable.

20

Lasers in Medicine

A laser can quickly punch a hole through a thick steel plate. Obviously, it could also be used as an absolutely devastating weapon.

What might not be so obvious is that lasers can also be used for healing. More and more medical applications for lasers are being developed. A finely focused laser beam can perform more delicate surgery than the sharpest scalpel. Many surgical techniques have been greatly simplified by the use of lasers. Some procedures that were formerly impossible can be handled with laser equipment.

In this chapter we will just take a quick glance at a few representative medical applications of the laser.

UNBLOCKING ARTERIES

A major cause of heart attacks and similar ailments is a blockage in one or more of the arteries supplying the heart.

In some cases, the blockage or part of the artery can be removed or bypassed. This is complicated and often dangerous. In some cases the blockage is totally inaccessible to ordinary surgical procedures.

A new approach, using lasers and fiber optics offers new hope. A tiny optical fiber tipped with a metal cap is inserted into a blood vessel near the skin. The fiber is then worked up through the veins like a plumber's snake until it hits the blockage. Then a laser beam is conducted through the optical fiber, heating the metal cap to about 750°. This is sufficient to melt through the fibrous, waxy deposits blocking the artery. The intense heat is very localized, and applied in brief pulses, so no damage is done to surrounding, healthy tissues.

EYE SURGERY

Operating on a human eye has always been a delicate procedure at best. A laser beam can cut much more finely and accurately than a scalpel. Often no incision needs to be made at all, since the eye is designed to admit light. The intensely focused laser beam can burn a tiny hole at any desired point in the eye with great accuracy.

These special properties of the laser have produced many advances in eye surgery. Perhaps the most important of these applications involves cases of retinal detachment. In this condition the retina, the photosensitive tissue which contains the visual receptors (rods and cones), rips or tears in some spot. The retina may continue to weaken until it detaches completely from the back of the eye. People with extreme near-sightedness or who were premature as infants often have retinas that have been weakened in some way that could lead to a detachment. In the past, this sort of retinal degeneration used to lead to almost certain blindness, but can now be treated easily and effectively with the laser.

In this operation, pinpoint bursts of laser light are used to burn tiny holes into the retina. The surgeon burns a series of several hundred of these holes which surround the lesion, thus sealing it off. The burned spots develop a tough scar tissue which helps to hold the weakened retina in place. The whole procedure is not unlike spot-welding.

Another major application is in the treatment of diabetic blindness. Here the small blood vessels that carry oxygen to the retina begin to sprout and grow. The new blood vessels are very fragile and may rupture, resulting in hemorrhaging. They may also produce traction on the retina which could lead to a detachment. The surgeon fires the laser at the affected area in a checkerboard pattern in a procedure known as "pattern-bombing." For some reason not yet understood, this turns off the growth stimulus to the blood vessels and prevents further deterioration which could result in blindness.

These operations can be done quickly and easily on an out-patient basis. They are usually painless and, because of the almost microscopic size of the holes, have almost no impact on visual acuity. In fact, many people experience an improvement in vision because of the reduction in retinal swelling due to the tear.

Despite popular assumptions, lasers are not used to remove cataracts. However, the back of the eye's lens capsule sometimes starts to cloud up a year or two after cataract surgery. If this happens, a laser can be used to clear up the cloudiness.

DRUG DOSAGES

Lasers even have medical applications outside of surgery.

One of the trickiest aspects of medical diagnosis is determining the proper dosage of a drug for a given patient. Everyone's system is a little bit different, and different people don't always react to a given drug in exactly the same way. An insufficient dose for one patient could conceivably be a dangerous overdose for another patient.

To complicate matters even further, the difference between an ineffective underdose and a dangerous overdose is often a very small range.

In the past, it has been impossible to locate and measure very minute concentrations of a drug in any bodily fluid, such as blood or urine. The doctor essentially had to guess the dose (based on statistical averages), and then increase or decrease the dose only if the patient showed signs of an incorrect dosage. While many doctors were very good at making these dosage estimates, the procedure was always a little uncertain at best.

A new technique using lasers can be used to detect drugs in concentrations as small as one part in a trillion. This new technique takes advantage of the monochromaticity of laser light. All atoms and molecules are limited in the numbers of colors they can absorb and emit. When the mono-colored laser beam passes through the examined fluid, a reaction will be stimulated in certain types of molecules, but not in others. Each drug has its own unique reaction "signature," so the presence and strength of any specific substance (drug) can be detected and precisely measured.

Similar procedures can be used in the chemical laboratory, which brings us to the next chapter.

Lasers in the Laboratory

The laser can be a valuable research tool for the scientist in the laboratory. Of course, its value in the study of the nature of light is obvious. But lasers can be useful in almost any field of science.

LASER-INDUCED CHEMISTRY

Lasers have a number of uses in the chemistry laboratory.

One use is in the identification and measurement of substances, as described in the drug dosages section of the last chapter.

When a beam of laser light enters a medium, the energy may be dissipated in a variety of ways. Some of the energy may be absorbed by certain substances. Unabsorbed energy may be dissipated as heat, or it may cause partial ionization.

When a substance absorbs the laser energy, it can respond in a number of ways. Molecules may be split, or ionized. Chemical changes can be induced by the heat, or, in some cases, by the wavelength of the light.

In certain photochemical reactions, a specific light wavelength will selectively excite certain molecular fragments, producing a limited chemical reaction. Generally the excitation wavelength will be in the ultraviolet region. Laser-induced chemistry has been considerably advanced by the fairly recent development of high-power ultraviolet lasers.

A laser beam aimed at a solid surface can initiate a chemical reaction at or above the surface. The chemical reaction induced can be due to thermal or photochemical processes, or a combination of the two.

LASER "COOLING"

Every element has its own unique spectrum. That is, each element reacts with specific wavelengths in a unique way. A laser can be used as a tool to help identify unknown substances.

The spectra of various elements is an important factor in many scientific processes. Unfortunately, the spectrum will always be somewhat blurred, broadened, or even shifted from its nominal value due to Doppler effects from the movement of the atoms.

In the past only ions (electrically charged atoms) could be held still enough to make any kind of precise measurement.

Recently, researchers have discovered a way to bring a stream of fast-moving atoms to a dead stop with a laser beam. A precisely tuned laser beam is aimed directly into an oncoming beam of free, neutral sodium atoms.

The atoms are slowed as if they were cooled to a point near absolute zero (the temperature at which all molecular movement stops). Currently this technique has only been successfully used on sodium atoms, but researchers are working on generalizing the process.

FREE-ELECTRON LASERS

In many scientific applications, an extremely short wavelength is required. Researchers have been striving for years to produce a laser in the X-ray region. They finally appear to be coming close to this goal.

A fairly new type of laser is the free-electron laser. This type of laser can produce either long or short wavelengths. Theoretically, a free-electron laser could be continuously tuned from microwaves on up into the X-ray region.

Basically, a free-electron laser efficiently converts electron beam energy into laser pulses. Instantaneous or peak output power can be as high as 80 megawatts.

One important potential application for free-electron lasers is energy containment in fusion energy production.

Laser Safety Hazards

The introduction and ever-increasing use of laser equipment has brought with it a number of new and hidden dangers not previously recognized. Information on these dangers to health must be disseminated to all personnel associated with laser equipment.

BIOLOGICAL EFFECTS OF LASER RADIATION

The body organ most sensitive to radiation is the eye. The ability of the eye to refract the near ultraviolet, visible, and near infrared radiation is the most important physiological characteristic contributing to the laser hazard. The energy density of the radiation incident on the cornea can be increased by as much as several hundred thousand times at the retina by the focusing power of the eye. This can create a major cataract hazard. If one looks directly into a laser beam, eyesight could be permanently destroyed.

Organs lying close to the skin are also susceptible to damage by laser radiation. Needless exposure of the skin to laser radiation should be avoided regardless of power density or energy density levels.

PROTECTIVE EQUIPMENT

Each particular laser system will require personnel protection specifically correlated to the operational parameters of the system.

Current laser filters are designed to filter a narrow band of wavelengths and usually offer protection for only one laser system.

Eye Protection

Filters can be worn in spectacle frames, goggles, and visors. They can also be fitted in windows and viewing ports. The application will dictate the weight and thickness of filter allowable. Four-millimeter-thick glass is generally considered maximum for most goggle applications.

All protective eye wear should be identified with the laser wavelength for which it is intended to be used, the optical density of that wavelength, and the visible light transmission.

Skin Protection

Should skin protection be required, heavy white cloth, such as a standard laboratory smock, reduces the exposure by at least a factor of 100 over the spectrum from the near infrared to the near ultraviolet. Gloves should be used to protect hands. Shielding and complete enclosure of the beam is preferable to personal protective devices and should be accomplished whenever possible.

GENERAL SAFETY PRECAUTIONS

There are some hazards common to all laser systems and other hazards that are specific to certain types.

Eye Exposure

Do not look into primary laser beams or at specular reflections. Special precautions should be taken to avoid accidental viewing of direct beams or specular reflections when power or energy densities exceed the PEL (permissible exposure level).

Beam Termination

Laser beams should be properly terminated by suitable targets or backstops that have no specular reflections and are fire resistant.

High Voltage

Many lasers employ high voltages. Proper safety precautions should be taken for high voltages as you would with any electronic equipment.

Toxic Chemicals

Toxic gases are often produced as a result of high-energy laser beams ionizing air or disintegrating the target.

Cryogenic Materials

There are many hazards associated with cryogenic liquids. Contact with the skin will produce a burn. Asbestos gloves should be worn when filling or pouring from Dewar and thermos bottles. The bottles should be protected by an outside container should the bottle implode.

Appendix

Atomic Elements

New kinds of lasers, using new kinds of materials for host and dopant, are being developed. Often these materials are identified by their atomic symbols. The tables in this appendix will be helpful in identifying the materials and in discerning some important facts about them that relate to their use in lasers and masers.

The periodic table of the elements (Table A-1) groups the elements according to their characteristics. The elements are identified by their atomic symbols.

The table of atomic symbols (Table A-2) lists the symbols in alphabetical order, along with the name of the element, its atomic number, and its atomic weight. This table will enable you to quickly identify the materials used in a laser.

The table of atomic elements (Table A-3) lists the elements themselves alphabetically, along with the atomic symbol, number, and weight.

Table A-1. Periodic Table of the Elements.

ATOMIC NUMBER — 1
ELEMENT SYMBOL — H
ATOMIC WEIGHT — 1.002

LIGHT METALS · HEAVY METALS · NONMETALS · INERT GASES

IA	IIA	IIIB	IVB	VB	VIB	VIIB	VIII	VIII	VIII	IB	IIB	IIIA	IVA	VA	VIA	VIIA	INERT GASES
1 H 1.008																	2 He 4.003
3 Li 6.94	4 Be 9.012											5 B 10.81	6 C 12.011	7 N 14.007	8 O 15.999	9 F 18.998	10 Ne 20.18
11 Na 22.990	12 Mg 24.305											13 Al 26.982	14 Si 28.086	15 P 30.974	16 S 32.06	17 Cl 35.453	18 Ar 39.95
19 K 39.102	20 Ca 40.08	21 Sc 44.956	22 Ti 47.90	23 V 50.942	24 Cr 51.996	25 Mn 54.938	26 Fe 55.847	27 Co 58.933	28 Ni 58.71	29 Cu 63.546	30 Zn 65.37	31 Ga 69.72	32 Ge 72.59	33 As 74.922	34 Se 78.96	35 Br 79.904	36 Kr 83.80
37 Rb 85.47	38 Sr 87.62	39 Y 88.905	40 Zr 91.22	41 Nb 92.906	42 Mo 95.94	43 ● Tc (99)	44 Ru 101.07	45 Rh 102.905	46 Pd 106.4	47 Ag 107.868	48 Cd 112.40	49 In 114.82	50 Sn 118.69	51 Sb 121.75	52 Te 127.60	53 I 126.904	54 Xe 131.30
55 Cs 132.905	56 Ba 137.34	57 La 138.91	72 Hf 178.49	73 Ta 180.948	74 W 183.85	75 Re 186.2	76 Os 190.2	77 Ir 192.2	78 Pt 195.09	79 Au 196.967	80 Hg 200.59	81 Tl 204.37	82 Pb 207.2	83 Bi 208.980	84 ● Po (210)	85 ● At (210)	86 ● Rn (222)
87 ● Fr (223)	88 ● Ra (226)	89 ● Ac (227)	104 ● Ku (257)														

LANTHANUM SERIES

58 Ce 140.12	59 Pr 140.907	60 Nd 144.24	61 Pm (147)	62 Sm 150.35	63 Eu 151.96	64 Gd 157.25	65 Tb 158.924	66 Dy 162.50	67 Ho 164.930	68 Er 167.26	69 Tm 168.934	70 Yb 173.04	71 Lu 174.97

ACTINIUM SERIES

90 ● Th 232.038	91 ● Pa (230	92 ● U 238.03	93 ● Np (237)	94 ● Pu (242)	95 ● Am (243)	96 ● Cm (247)	97 ● Bk (247)	98 ● Cf (249)	99 ● Es (254)	100 ● Fm (257)	101 ● Md (258)	102 ● No (255)	103 ● Lr (256)	105 ● (258-261)	106 ● (258-260)

● — INDICATES PRINCIPAL RADIOACTIVE ELEMENTS

Table A-2. Atomic Symbols

Symbol	Name	Atomic Number	Atomic Weight
A	Argon	18	39.948
Ac	Actinium	89	(227)
Ag	Silver	47	107.870
Al	Aluminum	13	26.98
Am	Ameridium	95	(243)
As	Arsenic	33	74.9216
At	Astatine	85	(210)
Au	Gold	79	196.967
B	Boron	5	10.811
Ba	Barium	56	137.34
Be	Beryllium	4	9.0122
Bi	Bismuth	83	208.980
Bk	Berkelium	97	(249)
Br	Bromine	35	79.909
C	Carbon	6	12.01115
Ca	Calcium	20	40.08
Cb	Columbium (Niobium)	41	92.906
Cd	Cadmium	48	112.40
Ce	Cerium	58	140.12
Cf	Californium	98	(251)
Cl	Chlorine	17	35.453
Cm	Curium	96	(247)
Co	Cobalt	27	58.9332
Cr	Chromium	24	51.996
Cs	Cesium	55	132.905
Cu	Copper	29	63.54
Dy	Dysprosium	66	162.50
E	Einsteinium	99	(254)
Er	Erbium	68	167.26
Eu	Europium	63	151.96
F	Fluorine	9	18.9984
Fe	Iron	26	55.847
Fm	Fermium	100	(253)
Fr	Francium	87	(223)
Ga	Gallium	31	69.72
Gd	Gadolium	64	157.25
Ge	Germanium	32	72.59
H	Hydrogen	1	1.00797
He	Helium	2	4.0026
Hf	Hafnium	72	178.49
Hg	Mercury	80	200.59
Ho	Holmium	67	164.930
I	Iodine	53	126.9044
In	Indium	49	114.82
Ir	Iridium	77	192.2
K	Potassium	19	39.102
Kr	Krypton	36	83.80
La	Lanthanum	57	138.91
Li	Lithium	3	6.939
Lu	Lutetium	71	174.97
Lw	Lawrencium	103	(257)
Md	Mendelevium	101	(256)
Mg	Magnesium	12	24.312

Symbol	Name	Atomic Number	Atomic Weight
Mn	Manganese	25	54. 9380
Mo	Molybdenum	42	95. 94
N	Nitrogen	7	14. 0067
Na	Sodium	11	22. 9898
Nd	Neodymium	60	144. 24
Ne	Neon	10	20. 183
Ni	Nickel	28	58. 71
* No	Nobelium	102	(254)
Np	Neptunium	93	(237)
O	Oxygen	8	15. 9994
Os	Osmium	76	190. 2
P	Phosphorus	15	30. 9738
Pa	Proteactinium	91	(231)
Pb	Lead	82	207. 19
Pd	Palladium	46	106. 4
Pm	Promethium	61	(147)
Po	Polonium	84	(210)
Pr	Praseodymium	59	140. 907
Pt	Platinum	78	195. 09
Pu	Plutonium	94	(242)
Ra	Radium	88	(226)
Rb	Rubidium	37	85. 47
Re	Rhenium	75	186. 2
Rh	Rhodium	45	102. 905
Rn	Radon	86	(222)
Ru	Ruthenium	44	101. 07
S	Sulfur	16	32. 0o4
Sb	Antimony	51	121. 75
Sc	Scandium	21	44. 956
Se	Selenium	34	78. 96
Si	Silicon	14	28. 086
Sm	Samarium	62	150. 35
Sn	Tin	50	118. 69
Sr	Strontium	38	87. 62
Ta	Tantalum	73	180. 948
Tb	Terbium	65	158. 924
Tc	Technetium	43	(99)
Te	Tellurium	52	127. 60
Th	Thorium	90	232. 038
Ti	Titanium	22	47. 90
Tl	Thallium	81	204. 37
Tm	Thulium	69	168. 934
U	Uranium	92	238. 03
V	Vanadium	23	50. 942
W	Tungsten	74	183. 85
Xe	Xenon	54	131. 30
Y	Yttrium	39	88. 905
Yb	Ytterbium	70	173. 04
Zn	Zinc	30	65. 37
Zr	Zirionium	40	91. 22

* Note: Element proposed but not confirmed.

Table A-3. Atomic Elements.

Name	Symbol	Atomic Number	Atomic Weight
Actinium	Ac	89	(227)
Aluminum	Al	13	26.98
Americium	Am	95	(243)
Antimony	Sb	51	121.75
Argon	A	18	39.948
Arsenic	As	33	74.9216
Astatine	At	85	(210)
Barium	Ba	56	137.34
Berkelium	Bk	97	(249)
Beryllium	Be	4	9.0122
Bismuth	Bi	83	208.980
Boron	B	5	10.811
Bromine	Br	35	79.909
Cadmium	Cd	48	112.40
Calcium	Ca	20	40.08
Californium	Cf	98	(251)
Carbon	C	6	12.01115
Cerium	Ce	58	140.12
Cesium	Cs	55	132.905
Chlorine	Cl	17	35.453
Chromium	Cr	24	51.996
Cobalt	Co	27	58.9332
Columbium (Niobium)	Cb	41	92.906
Copper	Cu	29	63.54
Curium	Cm	96	(247)
Dysprosium	Dy	66	162.50
Einsteinium	E	99	(254)
Erbium	Er	68	167.26
Europium	Eu	63	151.96
Fermium	Fm	100	(253)
Fluorine	F	9	18.9984
Francium	Fr	87	(223)
Gadolium	Gd	64	157.25
Gallium	Ga	31	69.72
Germanium	Ge	32	72.59
Gold	Au	79	196.967
Hafnium	Hf	72	178.49
Helium	He	2	4.0026
Holmium	Ho	67	164.930
Hydrogen	H	1	1.00797
Indium	In	49	114.82
Iodine	I	53	126.9044
Iridium	Ir	77	192.2
Iron	Fe	26	55.847
Krypton	Kr	36	83.80
Lanthanum	La	57	138.91
Lawrencium	Lw	103	(257)
Lead	Pb	82	207.19
Lithium	Li	3	6.939
Lutetium	Lu	71	174.97
Magnesium	Mg	12	24.312
Manganese	Mn	25	54.9380
Mercury	Hg	80	200.59
Mendelevium	Md	101	(256)

Name	Symbol	Atomic Number	Atomic Weight
Molybdenum	Mo	42	95.94
Neodymium	Nd	60	144.24
Neon	Ne	10	20.183
Neptunium	Np	93	(237)
Nickel	Ni	28	58.71
Niobium (See Columbium)			
Nitrogen	N	7	14.0067
* Nobelium	No	102	(254)
Osmium	Os	76	190.2
Oxygen	O	8	15.9994
Palladium	Pd	46	106.4
Phosphorus	P	15	30.9738
Platinum	Pt	78	195.09
Plutonium	Pu	94	(242)
Polonium	Po	84	(210)
Potassium	K	19	39.102
Praseodymium	Pr	59	140.907
Promethium	Pm	61	(147)
Protoactinium	Pa	91	(231)
Radium	Ra	88	(226)
Radon	Rn	86	(222)
Rhenium	Re	75	186.2
Rhodium	Rh	45	102.905
Rubidium	Rb	37	85.47
Ruthenium	Ru	44	101.07
Samarium	Sm	62	150.35
Scandium	Sc	21	44.956
Selenium	Se	34	78.96
Silicon	Si	14	28.086
Silver	Ag	47	107.870
Sodium	Na	11	22.9898
Strontium	Sr	38	87.62
Sulfur	S	16	32.064
Tantalum	Ta	73	180.948
Technetium	Tc	43	(99)
Tellurium	Te	52	127.60
Terbium	Tb	65	158.924
Thallium	Tl	81	204.37
Thorium	Th	90	232.038
Thulium	Tm	69	168.934
Tin	Sn	50	118.69
Titanium	Ti	22	47.90
Tungsten	W	74	183.85
Uranium	U	92	238.03
Vanadium	V	23	50.942
Xenon	Xe	54	131.30
Ytterbium	Yb	70	173.04
Yttrium	Y	39	88.905
Zinc	Zn	30	65.37
Zirconium	Zr	40	91.22

* Note: Element proposed but not confirmed.

Glossary

Most of the terms used in this book will be familiar to the person with a background in electronics. The terms that are used mainly with quantum electronics (masers and lasers) are explained in the text where they occur. This glossary is provided as an aid to the electronics reader who runs across an unfamiliar term in the text or in other maser and laser literature, also to the reader who comes from some field other than electronics.

amplification factor—The ratio of the change in anode voltage of an electron tube to a change in control-electrode voltage, when other tube voltages and currents are held constant.

anode—(1) The positive electrode through which the principal stream of electrons leaves the interelectrode space in an electron tube. Also called the plate. (2) The positive electrode of a battery or other electrochemical device.

audio frequency—A frequency that can be detected as a sound by the human ear.

avalanche—The process in which an electron or other charged particle accelerated by an electric field collides with and ionizes gas molecules, thereby releasing new electrons which in turn have more collisions.

avalanche breakdown—Nondestructive breakdown in a semiconductor diode when the electric field across the barrier region is strong enough so that current carriers collide with valence electrons to produce ionization and cumulative multiplication of carriers.

bandwidth—The width of a band of frequencies used for a specific purpose.

beam-power tube—An electron-beam tube which makes use of a directed electron beam to contribute to its power-handling capability.

beat note—The difference frequency obtained when two sinusoidal waves of different frequencies are combined.

bias—(1) A dc voltage applied to a transistor control electrode to establish the desired operating point. (2) The dc voltage applied between the control grid and cathode of an electron tube to establish the desired operating point. Also called grid bias.

bleeder current—Current drawn continuously from a voltage source to improve voltage regulation or to provide a voltage drop across a resistor.

bleeder resistor—A resistor connected across the output of a power supply (1) to protect equipment from excessive voltages if the load is removed or substantially reduced, and (2) to improve the voltage regulation, and to dissipate the charge remaining in filter capacitors when equipment is turned off.

buffer—(1) An amplifier used after an oscillator (or other critical stage) to isolate it from the load. Also called buffer amplifier and buffer stage. (2) The mode of operation of a digital computer that involves inter-equipment data transfer. (3) The final holding register on the computer's output.

carrier frequency—(1) The unmodulated frequency generated by a transmitter. (2) The average frequency of the transmitted wave when it is modulated by a symmetrical signal. Also called center frequency and resting frequency.

cat whisker—A small sharp-pointed wire used to make contact with a sensitive point on the surface of a semiconductor.

cathode—The primary source of electrons in an electron tube.

coaxial (coax)—Having one axis within another, as a coaxial cable, with a single cylindrical conductor suspended in the center of another conductor.

control grid—A grid usually placed between the cathode and an anode that serves to control the plate current of an electron tube.

corona—A bluish-purple glow surrounding a conductor (due to ionization of the surrounding air) when the voltage exceeds a certain critical value.

crystal—A natural or synthetic piezoelectric or semiconductor material.

degeneration—Negative feedback.

deionization—The recombination of ions with electrons in a glow or arc discharge to form neutral atoms and molecules.

deionization potential—The potential at which ionization of the gas in a gas-filled tube ceases and conduction stops.

detector—The stage in a receiver at which demodulation takes place. Also called demodulator.

discriminator—A circuit in which the magnitude and polarity of the output voltage depend on how an input signal differs from another signal or from a reference.

distortion—An undesired change in waveform.

distributed capacitance—The capacitance between adjacent turns in a coil or between adjacent conductors. Also called self-capacitance.

duplexer—A switching device used to permit alternate use of the same antenna for both transmitting and receiving.

duty cycle—(1) The product of pulse duration and pulse repetition frequency, which is equal to the time pulse power is on per second. (2) The ratio of pulse width to the time between corresponding pulses.

Edison effect—The emission of electrons from heated elements.

electrode—(1) A conducting element that performs one or more of the functions of emitting, collecting, or controlling the movements of electrons or ions in an electron tube, or the movements of electrons or holes in a semiconductor device. (2) A terminal or surface at which electricity passes from one material or medium to another.

emitter—A transistor electrode from which a flow of carriers enters the interelectrode region.

emitter follower—A grounded-collector transistor amplifier similar in operation to the electron tube cathode follower.

envelope—(1) The glass or metal housing of an electron tube. (2) A curve drawn to pass through the peaks of a graph showing the waveform of a modulated rf signal.

filament—A cathode made of resistance wire or ribbon, through which an electric current is passed to produce the high temperature required for emission of electrons in a thermionic electron tube. Also called directly heated cathode.

flashback voltage—The peak inverse voltage at which ionization occurs in a gas tube.

forward bias—A bias voltage applied to a pn junction with polarity such that a large current flows.

free-electron laser—A device which converts an electron energy beam into very powerful laser pulses.

free electrons—Electrons which are not bound to a particular atom.

ganging—A means of mechanically operating two or more controls with one control knob.

germanium—A brittle, grayish-white metallic element having semiconductor properties. Atomic number is 32.

getter—A special metal that is placed in an electron tube during manufacture and vaporized after the tube has been evacuated.

grid—An electrode located between the cathode and anode of an electron tube and having one or more openings through which electrons or ions can pass under certain conditions.

grid leak—A resistor connected in the grid circuit of an electron tube to provide a discharge path for the grid capacitor and to limit the accumulation of charge on the grid.

harmonic—An integral multiple of a fundamental frequency. The frequency of the second harmonic is twice the frequency of the fundamental or first harmonic.

hole injection—Holes produced in n-type semiconductor material when a voltage is applied to a sharp metal point in contact with the surface of the material.

hum—(1) An electrical disturbance occurring at the power supply frequency or its harmonics. (2) A sound produced by an iron core of a transformer due to loose laminations or magnetostrictive effect.

iconoscope—A television camera tube in which a beam of high-velocity electrons scans a photoemissive mosaic that is capable of storing an electric charge pattern corresponding to an optical image focused on the mosaic.

interelectrode capacitance—The capacitance between one electrode of an electron tube and the next electrode on the anode side.

ionization—A process by which a neutral atom or molecule loses or gains electrons, thereby acquiring a net charge and becoming an ion.

junction—(1) A region of transition between two different semiconducting regions in a semiconductor device, such as a pn junction. (2) A fitting used to join a branch waveguide at an angle to a main waveguide, as in a T-junction.

leakage inductance—Self-inductance in a transformer due to leakage flux.

leakage reactance—Inductive reactance in a transformer due to leakage flux that links only the primary winding.

low-pass filter—A filter which passes alternating currents below a given cutoff frequency and attenuates or blocks all other currents.

microminiaturization—Miniaturization involving the design and construction of equipment with dimensions smaller than that with subminiature techniques, generally done by integrating circuit elements with the device itself.

microphonics—Noise caused by the vibration of the elements of an electron tube.

mixer—An electron tube or semiconductor device and circuit used to combine the incoming and local oscillator frequencies to produce an intermediate frequency.

mho—The unit of conductance or admittance. It is the reciprocal of the ohm.

modulation—The process of varying the amplitude, frequency, or phase of a carrier wave in accordance with other signals to convey intelligence.

modulator—The circuit which provides the signal to effect modulation.

multiplexing—The transmission of two or more signals on the same frequency during the same period of time normally required for the transmission of a single signal.

multivibrator—A form of relaxation oscillator comprising two stages so coupled that the input of each stage is derived from the output of the other.

negative resistance—A resistance such that when the current through it increases, the voltage drop across the resistance decreases.

neutralizing capacitor—A capacitor, usually variable, used in an anode-to-grid feedback path in an amplifier circuit for neutralization purposes.

photochemical reaction—A chemical process induced by exposure to light (usually at a specific wavelength).

photoelectric emission—The emission of electrons by certain materials upon exposure to radiation in and near the visible region of the spectrum.

piezoelectric—Having the ability to generate a voltage when mechanical force is applied, or to produce a mechanical force when a voltage is applied.

plate—The positive electrode through which a principal stream of electrons leaves the interelectrode space in an electron tube. Also called anode.

power pack—A power supply unit.

primary electron—An electron emitted directly by a material rather than as a result of a collision.

relaxation oscillator—An oscillator which generates a non-sinusoidal waveform by gradually storing electric energy, then quickly releasing that energy.

reverse bias—A bias voltage applied to a pn junction with polarity such that little or no current flows.

ripple—The ac component in the output of a dc power supply due to incomplete filtering.

saturation current—The maximum possible current that can be obtained as the voltage applied to a device is increased.

saturation voltage—The minimum voltage needed to produce saturation current.

screen grid—A grid placed between a control grid and an anode of an electron tube and usually maintained at a fixed positive potential.

secondary electron—(1) An electron emitted as a result of bombardment of a material by an incident electron. (2) An electron whose motion is due to a transfer of momentum from primary radiation.

semiconductor—A material whose resistivity is between that of insulators and conductors.

semiconductor device—An electron device in which conduction takes place within a semiconductor.

shot effect—Noise in an electron tube due to the variation in the number and velocity of electrons emitted by the cathode. Also called shot noise.

space charge—The net electric charge distributed throughout a volume or space, such as the cloud of electrons in the space near the cathode of a thermionic electron tube or phototube.

stage—A circuit containing a single section of an electron tube or equivalent device, or two or more similar sections connected in parallel, push-pull, or push-push operation. It includes all parts connected between the control-grid input terminal of the device and the input terminal of the next adjacent stage.

stray capacitance—Capacitance between wires, between wires and the chassis, or between components and the chassis of electronic equipment.

suppressor grid—A grid placed between two positive electrodes.

swamping resistor—A resistor connected in the emitter circuit of a transistor to reduce the temperature effect on the emitter-base junction resistance.

tank circuit—A resonant circuit in which the applied voltage is connected across a parallel circuit formed by a capacitor and an inductor. Also called tank.

thermal agitation—Random movements of the free electrons in a conductor.

thermionic emission—The liberation of electrons of ions from a solid or liquid as result of heat.

thyratron—A hot-cathode gas tube in which one or more control electrodes initiate but do not limit the anode current except under certain operating conditions.

transconductance—An electron-tube rating, equal to the change in anode current divided by the change in control-grid voltage.

transit time—The time required for an electron or other charge carrier to travel between two electrodes in an electron tube or transistor.

trigger—(1) To initiate circuit action by applying a pulse to a trigger circuit. (2) The pulse used to initiate the action in a trigger circuit.

trimmer—A variable capacitor, potentiometer, or inductor in tuning circuits for alignment purposes.

velocity modulation—Modulation in which the velocity of an electron beam is controlled to produce a grouping or bunching of electrons.

Index

M

N

O

P

Other Bestsellers From TAB

☐ **CHIP TALK: PROJECTS IN SPEECH SYNTHESIS—Dave Prochnow**

Speech synthesis is one of today's most fascinating new technologies. And now it's one you can explore easily and inexpensively with the help of this excellent introduction to the principles of speech synthesis, complete with step-by-step instructions for constructing a variety of working synthesizers including both stand-alone and computer-interfaced units—at a fraction of the cost of comparable commercial units. 224 pp., 131 illus.
Paper $14.95 Hard $24.95
Book No. 2812

☐ **THE ELECTRONICS MANUAL TO INDUSTRIAL AUTOMATION—G. Randy Slone**

You'll go through a complete industrial training course on how automation works, its industrial electronic application, and how to repair and service automation equipment. Best of all, this book offers information that is understandable not only to those with engineering degrees, but to any industrial professional who must live with automation, maintain it, and effectively utilize it in the workplace. This guide will help you make intelligent decisions regarding the purchase and implementation of automation equipment and bring you up to date with all the most modern types of automation systems. 352 pp., 178 illus.
Vinyl $17.95 Book No. 2762

☐ **BUILD YOUR OWN WORKING ROBOT—THE SECOND GENERATION—David L. Heiserman**

Are you bored with ordinary, every-day types of electronic project? Ready for a real challenge? Meet Buster . . . an amazing, personal robot that you can build yourself. Using a step-by-step approach, the author breaks down the construction of a complex robot system into easy and enjoyable subsystems: You'll get instructions on the two modes Buster operates under—his independent one in which he uses his blunder response to run freely without operator control . . . and dependent mode in which he is in the control of the person using the Operator's Control Panel. 140 pp., 37 illus.
Paper $12.95 Hard $18.95
Book No. 2781

☐ **62 HOME REMOTE CONTROL AND AUTOMATION PROJECTS—Delton T. Horn**

Here are 62 different remote control and automation units that anyone can build. All parts and components are readily obtainable and the cost is only a fraction of that charged for ready-built devices in electronics stores . . . some of the devices can't be purchased commercially at any price! Just think how you can impress your friends with your high-tech know-how while making your home safer and more convenient. 294 pp., 222 illus.
Paper $12.95 Hard $18.95
Book No. 2735

Other Bestsellers From TAB

☐ **PARTICLES IN NATURE: THE CHRONOLOGICAL DISCOVERY OF THE NEW PHYSICS—John H. Mauldin**

If you're interested in physics, science, astronomy, or natural history, you will find this presentation of the particle view of nature fascinating, informative, and entertaining. John Mauldin has done what few other science writers have been able to accomplish . . . he's reduced the complex concepts of particle physics to understandable terms and ideas. This enlightening guide makes particle physics seem less abstract—it shows significant spin-offs that have resulted from research done, and gives a glimpse of future research that promises to be of practical value to everyone. 304 pp., 169 illus. 16 Full-Color Pages, 14 Pages of Black & White Photos. Large Format (7″ × 10″.

Paper $16.95 Hard $23.95
Book No. 2615

☐ **333 *MORE* SCIENCE TRICKS AND EXPERIMENTS—Robert J. Brown**

Here's an ideal way to introduce youngsters of all ages to the wonders and complexities of science . . . a collection of tricks and experiments that can be accomplished with ordinary tools and materials. You can demonstrate that air is "elastic," perform hydrotropism, geotropism, plant force, phototropism, or psychological tricks . . . or perform any one or more than 300 fascinating experiments. 240 pp., 189 illus.

Paper $10.95 Hard $15.95
Book No. 1835

☐ **TIME GATE: HURTLING BACKWARD THROUGH HISTORY—Charles R. Pellegrino**

Taking a new approach to time travel, this totally fascinating history of life on Earth transports you backward from today's modern world through the very beginnings of man's existence. Interwoven with stories and anecdotes, and illustrated with exceptional drawings and photographs, this is history as it should always have been written! It will have you spellbound from first page to last! 288 pp., 142 illus. 7″ × 10″.

Paper $16.95 Book No. 1863

☐ **333 SCIENCE TRICKS AND EXPERIMENTS—Robert J. Brown**

Here is a delightful collection of experiments and "tricks" that demonstrate a variety of well-known, and not so well-known, scientific principles and illusions. Find tricks based on inertia, momentum, and sound projects based on biology, water surface tension, gravity and centrifugal force, heat, and light. Every experiment is easy to understand and construct . . . using ordinary household items. 208 pp., 189 illus.

Paper $9.95 Hard $15.95
Book No. 1825

☐ **VIOLENT WEATHER: HURRICANES, TORNADOES AND STORMS—Stan Gibilisco**

What causes violent storms at sea? Hurricane force winds? Hail the size of grapefruit? Blinding snowstorms and tornadoes? The answers to all these and many more